Mathias Roth

On Paralysis in Infancy, Childhood, Youth and on the Prevention and Treatment of Paralytic Deformities

Mathias Roth

On Paralysis in Infancy, Childhood, Youth and on the Prevention and Treatment of Paralytic Deformities

ISBN/EAN: 9783337372927

Printed in Europe, USA, Canada, Australia, Japan

Cover: Foto ©berggeist007 / pixelio.de

More available books at **www.hansebooks.com**

ON PARALYSIS

IN

INFANCY, CHILDHOOD, AND YOUTH,

AND ON THE

PREVENTION AND TREATMENT

OF

PARALYTIC DEFORMITIES.

BY

DR. MATHIAS ROTH,

AUTHOR OF SEVERAL WORKS ON THE MOVEMENT-CURE, LATERAL CURVATURE, ETC. ETC.

WITH FORTY-FIVE ENGRAVINGS.

LONDON:
GROOMBRIDGE AND SONS,
5, PATERNOSTER ROW.

1869.

Dedicated to

Dr. R. E. DUDGEON,

AS A

TOKEN OF FRIENDSHIP AND GRATITUDE,

BY

THE AUTHOR.

———◆———

My dear Dr. Dudgeon,

In 1849 the Austrians, defeated in the battle near Tallya, evacuated Kassa* in the greatest hurry, leaving from 800 to 1000 sick and wounded Austrian and Hungarian soldiers behind them to the mercy of the Hungarian authorities, at whose request the professional men of the little town took charge of these unhappy soldiers, who had neither medicines, food, nor any other comfort. After having attended to a part of these wounded for a short time, I had to proceed to the seat of government, to obtain the necessary power for acting as Inspector-General of Hospitals.

It happened by chance that at the same time the government wanted to send somebody abroad, and I was entrusted with a confidential mission. Being obliged to pass the lines

* *Read* Kasha.

of the enemy, I was arrested, imprisoned, and while the political prosecution instituted against me was still pending the Russian invasion crushed the Hungarian cause.

In consequence of these events, after nine years of successful practice, I was obliged, as an exile, without means and introductions, to recommence, in a foreign country, a medical career. At that time you and several of my colleagues received the exile very kindly, encouraged me to remain in this country, and when, at the advice of an intelligent layman, I had chosen a speciality, you were the first to give me a proof of your confidence by placing some of your patients under my care. Your example was followed by other colleagues, and I am happy to have now an opportunity of publicly expressing to you and to them my gratitude for the kindness, the friendship, and the confidence bestowed during the past twenty years on

Yours, very sincerely,

M. ROTH.

PREFACE.

The study of paralytic affections in infancy and childhood, and of the deformities they cause, has been and is still much neglected. Many young practitioners begin their professional life without having had an opportunity of watching the treatment of these complaints, which require medicinal, surgical, dynamical, and hygienic means, as well as a good amount of patience and perseverance.

This treatise, although very incomplete, gives a short outline of some forms of the disease, and of the treatment which has hitherto proved tolerably successful. Thus, I hope it will be of some use to my younger professional brethren, and to such practitioners as have no time to read monographs written in foreign languages, and other works on a class of complaints which, neglected at, or soon after their invasion, have often an injurious effect for life.

I am convinced, and wish others may share my conviction, that a large number, probably the majority of deformities existing amongst infants, children and youths, can be prevented by a judicious management at the right period, and that many a paralysed or deformed patient can be saved—

1. From being victimised by that class of orthopædic or other surgeons, whose panacea is but the screw and the knife;

2. Or from falling a prey to the professional rubber and bonesetter, or the so-called *consulting* orthopædic instrument makers and electric machine manufacturers, who shun medical interference, and sell indiscriminately their instruments to the ignorant, just as the druggist sells his wares to poor patients.

Many of the suggestions regarding the treatment will be found useful, not for the young only, but also in cases where the age of the patient is more advanced, and which at present are treated by medicines only, or left entirely to their fate.

TABLE OF CONTENTS.

No.		Page
1.	Conditions under which Paralysis arises	1
2.	Additional Causes of Paralysis	2
3.	Paralysis frequently a Symptom of other Complaints	2
4.	Division of Paralysis with regard to the Nervous System	3
5.	The (Atrophic Localising) Infantile Paralysis	3
6.	Reasons for suggesting the names Atrophic and Localising	3
7.	The Characteristics of the Atrophic and Localising Paralysis in Infancy	4
	Violent or slight Feverish State precedes the Paralysis	4
	Convulsions and painful Contractions not characteristic	5
	Infantile Paralysis occurs also without any Fever or Convulsion	5
	Loss of Motion in the Trunk, Neck, and four Extremities, but not in the Face	5
	Two Cases of the Invasion of this Paralysis	6
	Forms of Localising Infantile Paralysis	7
	The Localisation of this form of Paralysis	7
	Origin of Deformities	8
	Origin of Deformities in the Lower Extremities	8
	Condition of Sensation, Reflex Action, and Electric Contractility	9
	Atrophy of the Paralysed Parts	9
	The Temperature of the Paralysed Parts	10
	Comparative Table of Temperature in the Atrophic Localising Paralysis, and those forms of Infantile Paralysis which are consequences of Cerebral Causes	11
	Cerebral Paralysis of Children	12
	Fatty Degeneration not an essential Symptom of this Complaint	13
	Predisposing Causes	13
	Paralysie Myogenique	14
	Paralysis, according to report, caused by Difficult Dentition	15

No.		Page
	Constitutional State preceding Infantile Paralysis	15
	Pathological Anatomy	15
	Results of Laborde's Autopsies	17
	Account of the Case	19
	Autopsy	20
	Granular Degeneration	20
	Changes of Muscular Tissue in the Atrophied Parts	23
	Fatty Degeneration	24
8.	Paralytic Deformities	26
9.	Talipes Varus Equinus	27
10.	Talipes Equinus	28
11.	Talipes Valgus	29
12.	Talipes Calcaneus	30
13.	Congenital Club-foot	31
	Pes Valgus Congenitalis	31
	Pes Equinus Congenitalis, slightly complicated with Varus	31
	Pes Varus Equinus Congenitalis	32
	Talipes Calcaneus Congenitalis	33
	Pes Varus Congenitalis	34
	Pes Valgus Congenitalis	35
	The Paralytic Deformities of the Knee-Joint	35
14.	Genu Incurvatum (Valgum) Paralyticum	35
15.	Genu Excurvatum (Varum) Paralyticum	36
16.	Genu Recurvatum	36
17.	Contracted Knee	37
18.	Paralytic Affections of the Hip	37
	Contracted Hip	37
19.	The Relaxed Hip	38
	Paralytic Affections of the Spine	38
20.	The Incurvation of the Spine, or Paralytic Lordosis	38
21.	The Lateral Curvature of the Spine	39
22.	The Paralytic Kyphotic Curvature	40
23.	The Paralytic Deformity of the Neck	40
	The Deformity of the Shoulder	40
24.	Paralytic Deformities of the Elbow and Wrist-Joint	40
25.	General Paralysis	41
26.	Diagnosis	43
27.	Diagnostic Symptoms in Cerebral Infantile Paralysis	45
28.	Facial Paralysis, or Paralysis of Portio Dura of the Seventh Nerve (this also combined with Cerebral Hemiplegia)	46
29.	Diagnosis of Paralysis from Spinal Congestion	47
30.	The Difference of Symptoms in Spinal Congestion	47
31.	Progressive Atrophic Muscular Paralysis	48
32.	Diagnosis of Myelitic Paraplegia	49

No.		Page
33.	Paralysis after Spinal and Cerebral Hæmorrhage	49
34.	Duchenne's Pseudo-Hypertrophic Infantile Paralysis, or Infantile Paralysis with apparent Muscular Hypertrophy	49
35.	Pseudo-Hypertrophic Paralysis	50
	Hillier's Case	51
	Foster's Case	51
36.	Traumatic Paralysis	52
37.	Arthritic and Rheumatic Paralysis	53
38.	Diphtheritic Paralysis	53
39.	Rickety Paralysis	53
40.	Paralysis through Pott's Disease	54
41.	Paralysis after Chorea	54
42.	Paralysis from Epilepsy and Convulsions	54
43.	Paralysis produced by Metallic Cosmetics	55
44.	Mercurial Paralysis	56
45.	Paralysis from Bad Habits and too early and frequent Abuse of the Sexual Functions	56
46.	Temporary Paralysis	57
47.	Paralysis caused by Cancer or Tumours in the Brain or Spine	58
48.	Congenital Paralysis	58
49.	Hip Disease	58
50.	Indications for the Treatment of Paralysis, and of the Deformities caused thereby, in Infancy, Childhood, and Youth	59
	Injury often produced by the improper use of Orthopædic Instruments and Tenotomy	60
51.	Indications for the Treatment of Localised Infantile Paralysis	61
52.	Treatment of the (Atrophic and Localising) Infantile Paralysis	63
53.	Treatment after the Invasion of Paralysis	65
54.	Medical Treatment of various forms of Paralysis by the Physiological School	67
55.	Domestic and Popular Means recommended for the Treatment of Paralysed and Atrophic Limbs	67
56.	Frictions (recommended in Paralysis by the Ancients)	68
57.	Electricity	69
	Duchenne's remarks on Faradisation	70
	Cases of Atrophic Paralysis treated by Electricity	71
	Duchenne's Case treated by Faradisation, without amelioration	71
	How Electricity should be applied	73
	Abuse of Electricity weakens	74

Table of Contents.

No.		PAGE
58.	The Movement Cure, or Medical Gymnastics, Active and Passive Gymnastics	76
59.	Manipulations	76
	Frictions with Fatty and Oily Substances	77
	Works on Friction	78
60.	Gymnastic Treatment	79
	Dr. West on Exercise in Paralysis	80
	Dr. Radcliffe's remarks on the Advantages of Movements	81
61.	The various means for raising the Temperature of the Paralysed Parts	83
	American Machines for Rubbing	85
	Case of Paralysis successfully treated with Ice and Frictions	86
	Frictions	86
62.	On the Influence of Volition on the Paralysed Part, and how to stimulate the Power of Will	88
	The manner of rousing the Will, and increasing the Innervation	90
	Successive Stages of Volition	92
63.	To prevent the Contractions of the Joints, Retractions of the Muscles, and, consequently, the various Deformities caused by them	93
64.	Orthopædic Apparatus	94
65.	How Mechanical Supports should be made and used	96
	Designs of Apparatus made by Mathieu and Charrière	97
66.	Means for arresting the constantly progressing Atrophy	97
67.	To remove the Contractions of Muscles, Retraction of Tendons, and the Abnormal State of the Joints	98
68.	Tenotomy	98
69.	To retain the Normal Position of the Joints, and to prevent Curvatures of the Spine and Deformities of the Limbs	100
70.	That Supporting Apparatus, Crutches. Sticks, or Retention Contrivances, should be gradually removed, and only those which are indispensable retained	102
71.	The forms of Paralysis in which the Movement Cure, including all other suitable means, can cause an improvement	102
72.	Effect of the Movement Cure on the Paralysed	103
	The object of the Treatment by Movements	104
73.	Conditions desirable for Paralysed Patients while under the Treatment by Movements	104

CASES.

Case		Page
I.	Cerebral Paraplegia, with Lumbar Kyphosis, contracted Knees and Talipedes Equini	105
II.	Paraplegia, with Talipes Equinus	105
III.	Paralysis of one Leg, with Talipes Equinus Varus, and weakness of the other Foot	106
IV.	Paralysis of the Left Arm, and Paraplegia, with left Talipes Valgo-equinus	107
V.	Cerebral Hemiplegia, with deformity of the Spine, of the Left Arm, Hands, and Fingers, complicated with Epilepsy	108
VI.	Paralysis Incipiens	110
VII.	Paraplegia, with Pes Equinus, and one perfectly loose Hip	111
VIII.	Cerebral Paralysis of both Legs, with slight contraction of the Hips (Pedes Equini Vari), and Paralysis of one Arm, with contraction of the Elbow	112
IX.	Paraplegia, with Anæsthesia of one Leg and Hyperæsthesia of the other	113
X.	Local Paralysis of one Sterno-cleido-mastoideus, with slight Lateral Curvature	114
XI.	Local Paralysis, with a high degree of Limping	115

TABLE OF ENGRAVINGS.

	FIG.	PAGE
GRANULAR DEGENERATION.		
Changes of Muscular Tissue, first stage	1	23
,, ,, ,, second ,,	2	23
,, ,, ,, third ,,	3	23
,, ,, ,, fourth ,,	4	24
FATTY DEGENERATION.		
The Normal Muscular Fibres, with transversal stripes	5	25
The various stages of Fatty Degeneration.	6, 7, 8	25
,, ,, ,, ,,	9, 10, 11, 12	26
,, ,, ,, ,,	13, 14	27
Paralytic Club-feet	15	28
,, Equinus Varus, with Atrophy of the Right Leg	16	28
Talipes Equinus Paralyticus	17	29
,, Valgus ,,	18	30
,, Calcaneus ,,	19	30
,, Valgus Congenitalis	20	31
,, Equinus ,,	21	32
,, Varus Equinus ,,	22	32
,, Equinus Varus ,, (left and right)	23, 24	33
Talipedes Vari (more advanced stage)	25	33
Talipes Calcaneus Congenitalis	26, 27, 28	34
Talipedes Vari Congenitales	29, 30	35
The same, seen from the Sole	31, 32	35
Talipes Valgus Congenitalis	33	35
Genu Incurvatum	34	36
,, Recurvatum	35	37
Atrophy, with contracted Knees, one Genu Varum, the other Valgum	36	37
Atrophy, with contracted Hips, Knock-knees, and Club-feet	37	38
Paralytic Lordosis, Spinal Curvature	38	39
Atrophic Paralysis of the Forearm, with improved Club-hand	39	41
General Paralysis of all Limbs, with Atrophy, especially of Left Deltoideus, and *Polichinelle* Legs	40	42
Duchenne's Pseudo-Hypertrophic Paralysis	41	50
Mechanical supports	42, 43	95
,, ,,	44, 45	96

ON SOME FORMS OF PARALYSIS IN INFANCY, CHILDHOOD, AND YOUTH, AND ON THE PREVENTION AND CURE OF PARALYTIC DEFORMITIES.

EVERY variety of paralysis which occurs in adults can also occur at an earlier age; to speak of an infantile or juvenile paralysis as distinct from paralysis in adults is only a conventional term.

Conditions under which paralysis arises.

1. The conditions under which the various forms of paralysis arise are very numerous, and the most common and general causes are those which destroy that normal state which is essential either to the conditions of sensitive impressions to the brain, and to their perception by the mind, or to the conveyance of the will, or rather of the stimulus of the will through the nerves of motion. Pressure on the brain by a large effusion of blood or serum, by fracture and depression of the bones, by tumours, or by excessive fulness of the cerebral vessels, disorganisation of structure, in fact whatever prevents the free circulation of the blood through every part of the brain or spinal cord, may produce paralysis which will be either general or local, according to the extent of separate portions or tracts of the brain and spinal cord which are affected.

From this source also originates the difference of the principal varieties of paralysis. If a central part from which a sensitive nerve arises be seriously diseased or destroyed, there will be loss of sensation (anæsthesia, paralysis of sensation) in the part supplied by the sensitive nerve in question, but its natural power of motion will remain unaltered;

if a centre of origin for motor-nerve be diseased or destroyed, the part supplied therefrom will lose its motion, but retain its sensibility; if an amount of nervous matter, which is a centre both for sensitive and motor nerves, be diseased, loss of sensation and motion (or complete paralysis), loosening, or relaxing, will take place.

Additional causes of paralysis.

2. A morbid change in the blood, such as may be induced by typhoid fever, scarlatina, measles, diphtheria, or anæmia. Congestion and inflammation of the nervous centres, and of their surrounding membranes, organic changes in the structure of the brain, spinal cord, and the spine itself, convulsions, epilepsy, chorea, metallic and vegetable poisons, functional derangements of various organs which act by reflex action. Brown-Séquard asserts that cerebral paralysis may be reflex by an irritation from a remote morbid part of the brain being transmitted to another part whose function is to control muscular movement. Onanism and masturbation, as well as too premature and too frequent abuse of the sexual functions, exhaustion, mental and bodily overwork, mechanical and traumatic influences, are amongst the most frequent causes of the various species of paralysis.

Paralysis frequently a symptom of other complaints.

3. Paralysis cannot be said to be always an idiopathic disease, and should, therefore, be considered rather as a symptom of some disorder of the nervous system (often seated at a distance from the part where sensation, motion, and nutrition have been lost), or of a disease of any other organ acting by reflex action on the nerves. Congenital paralysis is often caused by deficiency or atrophy of central organs, and arrested intra-uterine development; the various hysterical, rheumatic, arthritic forms of paralysis, and those caused by congestion and inflammation in the brain and spine, and its membranes, are merely symptomatic.

Division of paralysis with regard to the diseased nervous organ.

4. For practical purposes relative to the seat of the diseased organ causing the paralysis, we speak of a central (cerebral or spinal) and peripheric paralysis; and with regard to the various functions which are affected, there is paralysis of sensation, motion, and nutrition (this last usually but erroneously called atrophy, because this term expresses only that the nutrition is diminished, but not that the nervous influence providing for the function of nutrition has ceased). Loss of motion and nutrition are often co-existent, or the latter is an effect of the former. Loss of motion and sensation are less frequently combined with loss of nutrition.

The (atrophic localising) infantile paralysis.

5. This disease is frequently called the essential paralysis of infancy, THE infantile paralysis (Laborde); spinal paralysis of children, spinale Kinderlähmung (Heine); fatty atrophic paralysis of infancy, paralysis atrophique graisseuse de l'enfance (Duchenne); paralysis myogenique (Bouchut); torpeur douloureux des jeunes enfants (Chassaignac); atrophic paralysis of children, with fatty degeneration, &c. These are some of the names given to a particular form of infantile paralysis, which has been and is still frequently confounded with a similar disease, caused either by metastasis during the critical stages of the febrile exanthemata of childhood, or by several of the pathogenetic influences mentioned in paragraph No. 2.

Reasons for suggesting the names atrophic and localising.

6. I believe that the words *atrophic* and *localising* will by naming two constant characteristics of this form of infantile paralysis prevent the confusion at present caused by the general term of infantile paralysis, because infants suffer just like adults from various forms of paralysis. As

this form of infantile paralysis causes the majority of shrivelled, lame, cold, half-dead limbs, contractions of joints, and all kind of deformities, it might be useful to mention it first, and to give an extract of all what is known about it. The principal sources are Heine's and Laborde's monographies, and I have preferred to consult the original German and French works instead of repeating quotations given by other authors. My own and original contributions refer more to the treatment of this complaint, which is a disease *sui generis*, and not to be mistaken for other forms of paralysis occurring in infancy.

The characteristics of the atrophic and localising paralysis in infancy are—

7. (*a*) That in the majority of cases it occurs in babies and infants at the age of from four months to five and six years.

From six months to two years Duchenne saw 37 out of 56 cases; West 27 out of 43 cases; Hillier 8 out of 12 cases; Laborde 16 out of 26 cases.

(*b*) Without premonitory symptoms a violent or slight feverish state precedes the paralysis. This fever lasts sometimes only from twelve to twenty-four hours, rarely more than ten days, and varies in intensity. In more severe cases the little patients are very restless, and the pulse rises to 140, 150; *there is no vomiting*, a symptom very frequently occurring in other fevers of children, and especially in cerebral affections; a kind of somnolence induces the medical man to believe that a comatose affection proceeding from the brain will be developed, but the short duration of the fever, the complete absence of cerebral symptoms, and the general paralysis of the limbs which at one stroke attacks all the limbs as well as the trunk and neck, shows the difference between this and cerebral paralysis, or other affections of the brain. Laborde has in 50 cases observed the fever 40 times; West in 32 cases 5 times. Often the fever is so slight and short that it is scarcely or not at all observed, and the mother or nurse find an infant paralysed in the morning, although, to all appearance, it was well when put to bed.

(c) Convulsions and painful contractions of the hand or knee have also been mentioned by Vogt and Brüniche amongst the symptoms preceding the infantile paralysis, but these contractions, which last for a longer or shorter time after some convulsions, are most intense at the beginning and gradually disappear. Although these contractions might exist with the infantile paralysis, they are not amongst the characteristic symptoms of this complaint.

(d) This form of infantile paralysis occurs also without any fever or convulsion having been previously observed. The infant which has begun to crawl, stand, or walk is in good health, not the slightest indisposition is noticed, and at once, without any known cause, the power of crawling, standing, or walking is lost. I have been lately consulted in two cases, both boys, one five, the other six years of age, of whom one when eleven months old could crawl; the other at the age of fourteen months could stand. The mother of the former observed one day that the child was dragging the left leg behind him instead of executing the movement of crawling. When the doctor examined the child the left leg in its whole length from the hip down was found paralysed. The other child was walking with his mother, and whilst with and near her fell suddenly, and from that moment the right fore-arm and hand and one leg were paralysed. I remember two cases where the paralysis occurred immediately after lancing the gums, and others during the period of dentition, or after intestinal irritation produced through indigested food or worms. Although these cases occurred after or without previous convulsions they were different from the atrophic and localising infantile paralysis which at its invasion is most sudden and intense both in extent and quality. The age of the little patient and the period of dentition has induced several authors to believe that the process of dentition is particularly predisposing to this affection.

(e) The loss of motion takes place in the trunk, neck, and four extremities, but not in the face; walking and standing are out of the question; in the worst cases even sitting is impossible, the spine, having lost its power, is unable to support the body, which falls to either side or forwards, according to the law of gravity, just as the head, equally powerless, drops in a similar manner.

(*f*) The following two cases will convey a good idea of the invasion of this paralysis. Laborde mentions the case of a strong and well-built girl, of three years of age, who walked when eighteen months old. On the 12th November, 1863, she was perfectly well, when depression of spirits, want of appetite, and violent fever, which lasted forty-eight hours, were noticed. On the fourth day (16th) the fever ceased, but the child could neither walk, stand, nor sit; all the limbs were found paralysed, and she was unable to prevent the head from sinking. Although pinching was slightly felt it did not cause any pain; tickling the feet did not produce any reflex movements in the legs; the urine passed at times involuntarily. It appears that croup had some time before preceded this attack of general paralysis; but the uvula being perfectly moveable, and the power of swallowing normal, the tone of the voice unaltered, the subsequent symptoms proved that it was not a diphtheric paralysis. The arms rapidly recovered their power of motion, but the legs remained paralysed. The left side is more affected, the muscles on this side are considerably atrophied, and even under the influence of electricity do not contract, and a tendency to talipes valgus is visible in the left foot.

The same author quotes another case, where, without any known cause, a healthy boy of two years, who had walked when eighteen months old, and never had any complaint, was seized with a violent fever which lasted twenty-four hours, and was followed by general paralysis of the trunk, neck, and all the limbs. A fortnight later the arms recovered their power, and two years after the spinal muscles had not entirely regained their action. The legs are still incompletely paralysed; the muscles and bones are to a great extent atrophied; and a tendency to varus equinus is seen in both feet; the legs during the unsteady walk are placed apart to preserve the equilibrium.

(g) *Forms of a. l. i. p.*—This infantile paralysis appears mostly under the form of paraplegia. In 30 cases observed by Laborde there were 21 of the above-mentioned form. Heine mentions that out of 192 cases of paralysis of children he had 158 of *this* infantile paralysis (which he called the spinal infantile paralysis). Of this number were—

37 Paraplegia,
34 Paralysis of one leg (which he called hemiplegia),
84 Partial paralysis,
2 Paralysis of a single arm,
1 Lordotic paralysis.

(h) *The localisation of this form of paralysis.* — The general paralysis which at the first stroke seizes the trunk, neck, and all the limbs, has a tendency to *localise itself*. The trunk and neck appear to recover first, then the upper extremities, and finally one leg, and even when localised in one limb there is a further tendency in the disease to settle itself in a group of muscles, affecting several joints or one only. The partial paralysis is sometimes only localised in the flexors or in the extensors, whilst adductors or abductors are not at all simultaneously affected. The localisation has taken place, according to observation, a fortnight, three weeks, one month, two to six months, and sometimes a year after the invasion of the disease. The paralysis of both upper extremities is seldom seen, but paralysis of one arm is more frequent. I have hitherto seen four cases of brachial paralysis in which the deltoideus was almost entirely wasted. As the atrophy is very considerable the skin appears to cover the shoulder-joint, and the form of the bones can be easily traced. Local atrophy of the deltoideus has also been observed after external injuries, a gangrenous ulcer after vaccination, continued suppuration after blisters. Setons and cauteries are also amongst the causes of a similar local atrophy, which will be easily distinguished from the paralytic atrophy by the previous history of the case and by the scars and unequal lines of the skin. Besides the deltoid, the flexors or extensors, the pronators and supinators of the arm, as well as the flexors of the metacarpus and the extensors of the fingers, are more or less paralysed and atrophied.

(*i*) *Origin of deformities.*—In course of time retraction of the muscles which have retained some power takes place and causes a deformity of the joint in the direction of their action—as all antagonistic action has ceased—thus club hands and fingers, claw fingers, in- and eversion of the wrist and fore-arms, and contractions of the elbow are formed. According to the severity of the loss of power and atrophy in various parts, combinations of several deformities are formed; the paralysed arm is in the beginning passively hanging down, its size diminishes in some cases slowly, and it is only when compared with the healthy arm that the arrest of development and the progress of atrophy is noticed.

(*k*) *Origin of deformities in the lower extremities.*—Both legs are frequently simultaneously attacked. In these cases the deformity in both varies; thus one foot is varus, the other valgus.

The groups of muscles most frequently paralysed in the legs are—

1. The flexors of the foot and extensors of the toes (extensor longus communis digitorum, extensor pollucis magnus, peroneus longus, and brevis lateralis).
2. The extensors of the foot (gastrocnemius, solæus), and flexors of the toes.
3. The extensors of the leg.
4. The flexors of the leg.
5. The adductors of the leg.

The single muscles most frequently paralysed are—

Extensor longus digitorum pedis.
Tibialis anticus.
Sterno-cleido mastoideus.
Deltoideus.

The mode by which the varieties of club-foot, deformities of the knee (genu incurvatum valgum, excurvatum varum, recurvatum and contractum), of the hip and spine are caused is partly by the paralysed muscles and partly by the incomplete or complete use of the muscles which have either partially or entirely retained their power.*

(*l*) *Condition of sensation, reflex action, and electric contractility.*—The loss of sensation is rarely complete but in the beginning, and in the more intense degrees of the disease sensation is less distinct. Hyperæsthesia has been sometimes seen. Reflex action is rarely entirely lost, often diminished; sometimes normal, although the paralysis may be general and most intense. Soon after the commencement of the paralysis the reflex action is usually entirely lost in those forms where not only the majority of the muscles of the lower extremities have lost their power, but also where the legs fall in a bent position outwards, the adductors of the thigh being perfectly paralysed.

Electricity does not produce any or merely a very slight contraction in the paralysed muscles; but this condition is not sufficient to determine whether the original disease is *the* atrophic localising infantile paralysis or a cerebral paralysis, because there are cases in which some muscles contract when stimulated by electricity, whilst others do not do so; some patients showing all the characteristics of this infantile paralysis still retain the power of electric contraction. Marshall Hall was the first to note the difference between a cerebral and spinal paralysis, or the electro-muscular contractility which also (according to Duchenne) is always absent in spinal paralysis. Lately, this presence or absence of electro-muscular contractility has been considered an absolute criterion for the diagnosis, although it is absurd to say that a case was first a cerebral and later a spinal paralysis, because this contractility was first preserved and afterwards lost. In cases where brain and spine are simultaneously the cause of paralysis, and where the power of electro-muscular contractions exists in some parts and is deficient in others, it would be difficult to give positive opinion on the subject of the forms of the two paralyses as long as we are restricted to judge by only one symptom.

(*m*) *Atrophy of the paralysed parts.* — The paralysed

* The description of the most frequent paralytic deformities will be found after the chapter on pathological anatomy.

parts are conspicuous by a particular flaccidity and softness of the muscles, by an atrophy of the muscular, ligamentous, and even osseous tissues, consequently by diminution of volume; the bones lose sometimes in length as well as in circumference. Heine observed the shoulder-blade and the knee-pan one third smaller in size on the paralysed side. I have also seen several cases where all the bones of the foot have been smaller, or where the length of the tibia and fibula was reduced by more than an inch, or the femur of the paralysed limb was shorter.

The data regarding the time when the atrophy begins are not yet sufficient for fixing the exact period. Some authors believe that it begins simultaneously with the paralysis, others only a few months after, in consequence of the loss of muscular activity. I believe that during what is called the acute stage of this paralysis the development is arrested, and the younger the child is the disproportion in the length and volume of the affected parts is greater; later, the affected parts, especially the bones, commence growing again, but never attain the development of the healthy part.

(*n*) *The temperature of the paralysed parts.*—The temperature of the paralysed limb is considerably lower; in the other form of cerebral paralysis the difference is rarely so great.

Heine has given, for this form of infantile paralysis, the following table of the temperature of the paralysed limbs in Reaumur's scale, to which I have added the reduction according to Fahrenheit.

Paralysis of one leg.

No.	Age.	Thigh.		Leg.		Sole of foot.	
		R.	F.	R.	F.	R.	F.
1	14 months	24	86	20	77	19	$74\frac{6}{5}$
2	26 ,,	23	$83\frac{6}{5}$	19	$74\frac{6}{5}$	$18\frac{1}{2}$	$73\frac{3}{5}$
3	36 ,,	20	77	17	$70\frac{2}{3}$	$16\frac{1}{2}$	$69\frac{1}{8}$

Paraplegia.

4	15 ,,	23	$83\frac{6}{5}$	22	$81\frac{4}{5}$	21	$79\frac{2}{5}$
5	27 ,,	22	$81\frac{4}{5}$	20	77	18	$72\frac{4}{5}$
6	40 ,,	21	$79\frac{2}{5}$	$17\frac{1}{2}$	$71\frac{3}{5}$	17	$70\frac{2}{5}$
7	52 ,,	19	$74\frac{6}{5}$	$17\frac{1}{2}$	$71\frac{3}{5}$	16	68

By comparing the temperature of the sole of the foot with that of the thigh, I find that the difference is greater in the paralysis of one leg where the maximum is $11\frac{2}{8}$ F. (in No. 1), the minimum (in No. 3) $7\frac{7}{8}$ degrees; while in *paraplegia* the maximum of difference is (in Nos. 5 and 6) 9 degrees, and the minimum in No. 4 is only $4\frac{4}{8}$.

Another interesting fact strikes me that the *average* of temperature diminishes in proportion to the more advanced age of the patient; thus, at the relative ages of 14 months (No. 1) and 15 months (No. 2), the amount of the average temperature in the paralysed limb is $79\frac{2}{8}$ and $81\frac{4}{8}$, while the average temperature at 36 months (No. 3) is $72\frac{1}{8}$, and at 52 months (No. 7) $71\frac{3}{8}$.

Another table by the same author shows the comparative temperature of the (a. l. i.) spinal paralysis and cerebral paralysis.

Comparative table of temperature in the atrophic localising paralysis and those forms of infantile paralysis which are consequences of cerebral causes.

14·0 Reaumur, $63\frac{4}{8}$ Fahrenheit, in the room.

Atrophic localising infantile paralysis.

No.	Age.	Form of paralysis.	Part measured.	Degree of temperature.	
				R.	F.
1	1½ years	Paraplegia	Armpit	27	$92\frac{6}{8}$
			Lumbar part of the back	26	$90\frac{4}{8}$
			Thigh	24	86
			Leg	20	77
			Sole of foot	18	$72\frac{4}{8}$
2	6 „	Paraplegia	Armpit	29	$97\frac{2}{8}$
			Lumbar part	26	$90\frac{4}{8}$
			Thigh	23	$83\frac{6}{8}$
			Leg	20	77
			Sole of foot	17	$70\frac{2}{8}$
3*	6 „	Paraplegia	Armpit	29	$97\frac{2}{8}$
			Lumbar part of spine	$26\frac{1}{2}$	$91\frac{4}{8}$
			Thigh	$18\frac{1}{2}$	$73\frac{5}{8}$
			Leg	17	$70\frac{2}{8}$
			Sole of foot	$14\frac{1}{2}$	$64\frac{4}{8}$

* This case is numbered 4 in Heine's table.

Cerebral paralysis of children.

No.	Age.	Form of paralysis.	Part measured.	Degree of temperature.	
				R.	F.
1	10 years	General paralysis. Perfect loss of motion and sensation	Armpit Spine, above the kyphois Thigh Leg Sole of foot	30 28½ 27 25½ 25	99⅘ 96⅛ 92⅖ 89⅗ 88⅖
2	10 ,,	Ditto.	Armpit Spine, above the kyphois Thigh Leg Sole of foot	29 28 27 26 26	97⅔ 95 92⅖ 90 90
3	7 ,,	Paraplegia.	Left side, armpit Middle of spine Thigh Leg Sole of foot	29 27 27 25 20	97⅔ 92⅖ 92⅖ 88¼ 77

By calculating the averages of the temperature in the preceding table, I find the following result:

	The temperature of			Difference of
	Thighs.	Legs.	Sole of foot.	thigh and foot.
In atrophic localising paralysis	81 6/48	... 74 36/48	68 46/48	12 8/48
In cerebral paralysis .	92 30/48	... 89 10/48	85 4/48	7 32/48
Difference	11 39/48	14 22/48	16 6/48	4 24/48

The last line shows the number of degrees by which the average temperature is *higher* in cerebral paralysis. In the last column to the right we may observe that the differences of temperature in the thigh and foot of the paralysed limb is four degrees and a half less in the cerebral paralysis, or in other words, that the average loss of warmth in the paralysed leg of the atrophic localising paralysis is four degrees and a half greater than in the cerebral paralysis.

The attention which has lately been paid to the changes of temperature in acute diseases has induced me to calculate the averages of temperature, especially as I believe that this considerable diminution of temperature, and the greater difference in the warmth of the various parts of the paralysed limb, forms a pathognomonic symptom of the a. l. i. paralysis.

Besides, the diminution of temperature shows itself not only by the thermometer but also by the discoloration of the skin, which varies in hue from the usual rose colour to blue, interspersed with deep-coloured spots. External influences of cold or heat easily produce bad sores, chilblains, and even gangrenous ulcers.

(*o*) *Fatty degeneration not an essential symptom of this complaint.*—Although the substitution of another especially fatty substance for the atrophied muscular fibres often takes place, it is not an essential symptom of the infantile paralysis, and this is another reason why this complaint should not be called "the fatty atrophic paralysis of infants." The granular degeneration or rather substitution for the muscular tissue is more frequent; the various degrees of this will be mentioned in the section of pathological anatomy.

(*p*) *Predisposing causes.*—To the predisposing causes of infantile paralysis, usually mentioned by various authors, belong the following :—a blow, a fall, compression, or a sudden pull or jerk on a part (commonly and wrongly attributed to the impatient nurse, especially when the paralysis has been localised in an upper extremity, and the deltoideus has been atrophied), or to a cold which is believed to be caused by a draught, exposure to damp, by sitting on a cold stone, damp grass, &c.

Although sometimes paralytic symptoms have been thus produced, these and some cerebral paralytic affections leave no trace behind, are only temporary, and quite different to the a. l. infantile paralysis, for which Rilliet and Barthes, Kennedy and others, have mistaken it, and Chaissagnac has called it the painful paralysis of children, whilst it is only temporary rheumatism or the effect of the mechanical cause. Pressure on a nerve continued sometimes for hours interferes with the free motion of the part affected, and the child is either prevented or frightened to move the limb because it is very painful; such a state of forced immobility cannot be called paralysis, even if it lasts for some time.

Bouchut's name of "Paralysie myogenique" is based on the inactivity of muscles caused by cold or rheumatism, and he attributes it to the neglect of suitably clothing infants and children, and exposing them in various ways to cold. Although infants and children are more exposed to cold about the neck and shoulders whilst in bed, the infantile paralysis is proportionately less observed in the upper extremities.

(*g*) *Two cases of paraplegia attributed to rolling on damp grass.*—I have seen a boy of eleven years of age who was in perfect health, and whilst on a tour in Wales sat down on the grass (perhaps after a fatiguing walk). It was supposed that this was the cause of a paraplegia brought on within a fortnight. From the account of the symptoms I thought a slight degree of myelitis preceded the loss of power in both limbs, which three or four months later were very considerably atrophied in their whole length, and the temperature lowered. Although he had been seen and treated by those specially engaged in the treatment of paralytic affections, and notwithstanding the prolonged application of electricity, passive manipulations, baths, &c., the atrophy of the muscles lasted, but the growth of the bones continued normally. In course of the last two years the os sacrum began to be deformed, and assumed a kyphotic appearance, which induced one medical man to believe that a heavy blow or fall was the cause of this paraplegia. My impression is that the consolidation of the os sacrum did not progress normally in consequence of the deranged power of assimilation. There was neither an abscess nor pain observed on the angular projection of the os sacrum, which a specialist promised to remove by electricity, which was in vain applied daily during at least six months.

I have also seen a well-built, healthy country girl, who, through sitting on the damp grass, had lost the use of both legs, and, notwithstanding her good health and the application of many kinds of treatment, she has only slightly improved. The atrophy in this case was not very considerable, and there was no tendency to localisation in either of the two cases.

(*r*) *Paralysis according to report caused by difficult dentition.*—During dentition, and especially during the eruption of the first and second tooth, paralysis has been frequently observed without previous convulsions or fever. I have seen several cases where, soon after the gums had been lanced (that is, the following morning), loss of motion had been remarked in a leg and foot, or hand. Amongst these were some children who had been born in India, and dentition was mentioned as the cause. But that, as well as worms, irritation of the stomach, and cold are most frequently *assumed* by parents and nurses to be the origin of a host of infantile diseases that may afterwards exist.

(*s*) *Constitutional state preceding infantile paralysis.*— Authors also differ regarding the constitutional state of the children who are affected by paralysis. Some say that the children are usually weak or sickly; others (Rilliet and Barthes) that they have been subject to eczema, impetigo, catarrh of the nose, and bronchial affections. Bouchut asserts that the children have usually been in perfect health and well developed. The majority of those I have seen had very good and well-developed heads, and the appearance of health and strength, and with the exception of the paralysed parts would have been considered fine children; only a few appeared to be weak and sickly. The majority of Heine's cases have been previously strong and healthy children.

(*t*) *Pathological anatomy.*—As the atrophic localising infantile paralysis is rarely fatal, there have been very few opportunities for post-mortem examinations; amongst these previously used to be mentioned four cases examined by Dr. Meryon, but these belong to a disease similar, if not identical, to muscular atrophy, with fatty degeneration, or to the so-called pseudo-hypertrophic paralysis of children, by which four brothers were affected.

Rilliet and Barthes mention two cases, but one of them seems to belong to another class of infantile paralysis, as there was neither atrophy nor fatty degeneration. The child died in consequence of pneumonia, and no anatomical lesions were found in the brain. In the second case no histological examination of the nervous centres was made. The same may be said of a case of Duchenne, who examined only the muscular tissues. Dr. Fliess has seen in a brachial paralysis a congestion of the meninges of the spinal marrow, at the part where the plexus brachialis originates. A spinal cord may have the normal colour, form, and consistence, and the pathological changes in the nervous tissues might be still very considerable, but the microscopic examination, as well as chromic acid and carmine, are required for preparing the parts to be examined; without these aids no autopsy of paralytic cases is of any use.

Berend has examined a boy who, when a year old, had a varus of the right foot, and died from cerebral affection when five years old. The atrophic right thigh and leg had $8\tfrac{1}{4}$ and 5 inches in circumference, while the corresponding healthy parts measured $11\tfrac{1}{2}$ and 8.

Much serum in the various parts of the brain; a firm gelatinous pseudo-membrane covers the whole space from the nervi optici to the corpora pyramidalia, and the anterior and posterior surface of the spinal marrow to the cauda equina. The paralysed muscles extremely pale, atrophic; the fibres very rarified, and much cellular tissue interspersed, but no fatty degeneration; the nerves not diminished in volume.

Longet's post-mortem examination of a paralytic club-foot in a girl of eight years.

The child died from smallpox six weeks after tenotomy was performed; the whole leg was very atrophic; the paleness of the paralysed muscles very considerable; the nerves of this leg thinner; the anterior roots of the lumbar and sacral nerves, forming the right ischiadieus, brown and ochre coloured, had scarcely a fourth of the diameter of the corresponding nerves on the healthy side; the posterior roots normal.

Dr. Hutin examined a case of paraplegia which, according to the symptoms, was caused by paralysis infantilis when seven years old, and where the patient's legs had been very seriously and extraordinarily deformed by rachitis; he was confined to his bed during his whole lifetime. He died at the age of forty-nine in Bicêtre.

In this case there was the highest degree of emaciation; brain healthy, ventricles, as well as the arachnoidea spinalis contained much serum; the spinal marrow above the eighth pair of dorsal nerves normal, but at this region harder and less in volume; the lower part, instead of having the usual enlargement, as in a healthy state, was diminished to the size of an ordinary quill, and was very hard. In the lumbar nerves the grey substance could not be discerned; the very thin nerves were yellow, dry, and hard; the vessels of the atrophied parts very small; the muscles of these parts had the appearance of white ligaments.

(*u*) *Results of Laborde's autopsies.*—A thorough post-mortem examination was made in the two following cases by Laborde:

An infant of eight months was attacked, after a short fever, by general infantile paralysis; the upper extremities soon recovered their power, but paraplegia continued; when two years old the atrophy was not considerable in the muscles of the legs, which contracted but slightly under electricity. The little patient was very intelligent, and no sign of a cerebral affection was observed. After a few months' treatment by local tonics, and the application of electricity, the limbs recovered some power of movement, when she was suddenly seized by some cerebral disease, with vomiting, strabismus, coma, and died.

With the exception of a well-marked congestive state of the meninges, the brain was normal; the spinal marrow, after the incision of the dura mater, showed normal volume and consistence; it was firm in its whole length, especially in the region of the lumbar enlargement, which, in children of this age, always shows a certain amount of rigidity. As soon as the healthy-looking pia mater was removed, an abnormal coloration of the anterior and lateral columns was visible, particularly the anterior, which, instead of their natural opacity, which was still seen in the posterior columns, were translucent, and of well-marked grey rose colour; the lateral columns had a similar appearance, but less distinctly pronounced. Transversal section of the spinal marrow at various heights showed the extent of the lesion affecting the whole substance of the anterior, and of the cortical portion only of the lateral columns; the surface, as well as the substance of all these parts, from the cerebral to the lumbar region, was greyish, transparent, and less firm than in the normal state.

The anterior and posterior roots did not appear altered to the naked eye.

The anterior and lateral columns showed, when magnified, 250—500 times, a remarkable (proliferation) increase of the elements of connective tissue, both of cellules and nuclei; the last measured $\frac{4-6}{1000}$ of a millimetre, and contained a scarcely visible little nucleus; the cells less abundant had a nucleus and little nucleus. These elementary parts were dispersed in a finely granular substance, composed of extremely thin fibrils. In those parts where these morbid tissues were more copious, the nervous tubes were scarcely to be recognised; they were, and appeared to be swollen and varicose, the amyloid bodies almost entirely wanting.

The small nervous cells, with the copious prolongations of the grey substance of the anterior horns, as well as the anatomic elements of the posterior columns were normal.

The capillary vessels filled with blood did not show any lesion, and neither prolification of nuclei, nor granular bodies were observed on their walls.

The microscopic examination of the spinal marrow (hardened in chromic acid) and of the transversal sections, coloured by an ammoniacal solution of carmine, confirmed the results mentioned above, and that the new production or increase of the connective tissue had taken place exclusively in the longitudinal tubes of the anterior and lateral columns, and that the horns of the grey substance remained in their healthy state.

The sciatic nerves examined in the fresh state, as well as after being immersed in chromic acid, did not show any alteration.

The paralysed muscles, although very minutely and

microscopically examined, did not show any lesion of their minute structure, only a slight diminution in the size, and a little discoloration of their fibres were observed.

The second thorough autopsy was made by the same author on a little boy of two years of age, who, when about one year old, had an attack of fever with repeated convulsions, during which he lost consciousness, became blue, and appeared to suffer extremely. From that time he could not walk; whether the upper limbs had been for the moment paralysed is not known, but even if this were the case he recovered their use at any rate very soon. The muscles of the lower extremities began to be rapidly atrophied, and his legs were considerably deformed. For this reason he was placed in the Children's Hospital, where, ten months later, he suffered from ophthalmia and measles, with double pneumonia which caused his death.

Account of the case.—The child looked most miserable; he was extremely emaciated, and was covered on the head and face with impetiginous crusts; both his legs were forcibly and constantly bent in the knees, which were turned outwards. The feet were stretched to such an extent that the heels touched the legs, whilst the dorsal surface of the foot was almost in a line with the tibia; the sole of the foot turned inwards and upwards, whilst the toes were slightly bent. Thus, a club foot, known as the worst degree of equinus varus, was formed. This deformity was greater on the left side, and was caused in the knee by the paralysis of the extensors, abductors of the leg, and by the retraction of the flexors, and in the foot by the paralysis and atrophy of the flexors and abductors of the foot by the predominant power and retraction of the gastrocnemii, and of the flexors of the toes; the resistance of these muscles was extreme as proved by the great tension of the tendo Achillis. The slightest attempt to bend the foot was extremely painful and impossible. At the upper extremities, the hands are turned back, the palms completely upwards; thus the wrist appears to have undergone a complete torsion; the fingers were slightly bent, the arm was forcibly stretched at the elbows, whilst the forearm was twisted outwards, and the anterior and interior surface was turned out or upwards. This deformity was less in the right arm and gave way to manipulations, whilst the replacing into a normal position of the left arm and hand was impossible in consequence of the very considerable retraction of the pronators and flexors of this side. Electro-contractility existed only in the gastrocnemii and in the upper limbs; it was very much diminished in the deltoideus and in the muscles of the back and arm.

The muscles on the anterior side of the leg were atrophied to such an extent that they appeared to have been entirely lost, and could not be felt. Dr. Duchenne believed that all these muscles had been transformed into adipose tissue; the voluntary movements had not been entirely lost, but they could be only done in a slight degree.

The slightest pinch was felt acutely and proved hyperæsthesia.

The little patient was very intelligent, understood all he was told, but the great weakness prevented him from speaking; he sat easily, and no spinal stiffness was observed. There was neither strabismus nor any other cerebral symptom; he passed urine in bed, whether involuntarily or only in consequence of his great weakness could not be ascertained.

Autopsy.—The brain in a normal state, slight bloody effusion of the meninges, and a very small quantity of reddish serum in the lateral ventricles. In the spinal cavity a thin layer of a turbid liquid, surrounding the spinal cord outside its membranes.

The spinal marrow increasing in volume from the top downwards, where a fluctuating tumour of the size of a pigeon's egg is found; a puncture in the spinal arachnoidea over the lumbar enlargement permits the escape of a yellowish turbid liquid, of a thick and almost gelatinous consistence (about 100 grammes in quantity); the tumour entirely disappears after the escape of the liquid.

The spinal pia mater is, in its whole extent, reddish; on its surface are very abundant vascular nets (*lascis*) and slight traces of white and milky stripes are seen which extend to the prolongation of the roots of the nerves.

The spinal marrow itself is in the whole length sufficiently firm, and appears externally normal; but when transversely dissected at various parts it seems gelatinous and transparent.

With the exception of the congestive state of the surrounding membranes, the anterior and posterior roots were found (under the microscope) in a healthy condition.

Although the histological examination of the marrow in its fresh state is rather difficult, it was proved that the small and large cells, and the tubes of the white substance, and of the prolongations of the grey substance were normal, and that there was no production of new tissue in the posterior column, and in the fundamental central substance; but this was the case in the anterior column; the longitudinal tubes were less numerous than in the healthy state; in some places they were entirely wanting, and those which remained were swollen, varicose, and as if broken in pieces. This was especially in the peripheric and superficial parts of the anterior column. The elementary parts of the nervous substance were altered in their structure, separated in fragments, and more or less granular little bodies were infiltrated within them; the capillary vessels of the pia mater, as well as of the peripheric nervous substance in contact with this membrane, show an increase of their nuclei, and their walls are studded over fully with a number of exudated corpuscles. This condition of the capillaries and nervous substance existed especially in the lumbar region of the cord, and was also very visible in the fibrous prolongations surrounding both roots of the spinal cord.

The microscopic examination of the spinal marrow saturated in chromic acid confirmed the previous observation in the peripheric parts, and in the walls of the capillary vessels.

The nerves of the arms and legs were found in a healthy state, with the exception of the left sciatic, in which the number of primitive nervous tubes was smaller in comparison with the other nerves, whilst the fibrillary elements of the connective tissues were abnormally multiplied.

The muscles of the upper extremities atrophied in various degrees; on the right side their size diminished, the colour normal, the deformity observed during life still present in a very slight degree. The changes on the left side are more marked; the deltoideus exists only in its anterior half, the rest is reduced to a thin, whitish little (bandelette) band of a fibrous appearance, and does not show the slightest trace of the normal muscular tissue. The super- and sub-spinalis scapulæ are partly atrophied; the other muscles of this limb have also lost in size, are very pale and discoloured, but still show traces of the fleshy structure.

In the lower extremities all these changes are still more considerable than those just described, and, notwithstanding they are atrophied to the highest degree, the skeleton of the muscles does not show the yellowish discoloration characteristic of the adipose or fatty degeneration. The details of this post-mortem show that in this case a previous and a recent pathologic process has taken place. The first left its traces in the changed structure of the nerve substance, whilst the recent and acute process shows its effects by the exudation and congestion, and accounts for the more or less retraction and atrophy of some muscles.*

Cornil found in a woman, forty-nine years old, whose paralysis dated from her second year, complete fatty degeneration of the muscles with atrophy of the primitive fibres, complete fatty degeneration of the nerves with atrophy of the nervous tubuli, and of the antero-posterior bundles of the cord, with production of amyloid corpuscles throughout its entire extent.

Drs. Bastian and Duchenne have found changes somewhat analogous to those of Laborde in the spinal cord after concussion and traumatic lesion. Bastian saw very extensive areas of degeneration consisting of atrophied nerve-fibres, new connective-tissue elements and granulation cor-

---

* The two post-mortems by Laborde are the best hitherto made, and there is no reason why Dr. Radcliffe (in Reynolds's *System of Medicine*) should assert that pathological anatomy has hitherto failed to show other than negative results.

puscles. Duchenne found the muscular symptoms consequent on traumatic lesion exactly the same as those in the atrophic paralysis of infancy.

Granular degeneration.

(*u*) *Changes of muscular tissue in the atrophied parts.**

In the first stage (fig. 1) traces of the transversal striæ are still visible, but the number of striæ is very much diminished, and separated by wider spaces, which are filled with molecular and opaque granules; a number of such granules cover also the remaining muscular bundles (fascicles) with their striæ. The granules are dissolved neither by alcohol nor ether, but by diluted acetic acid. This pathological state is found in the muscles, which in appearance are least changed, and have still retained some reddish fibres to be observed by the naked eye.

Fig. 1.

Fig. 2.

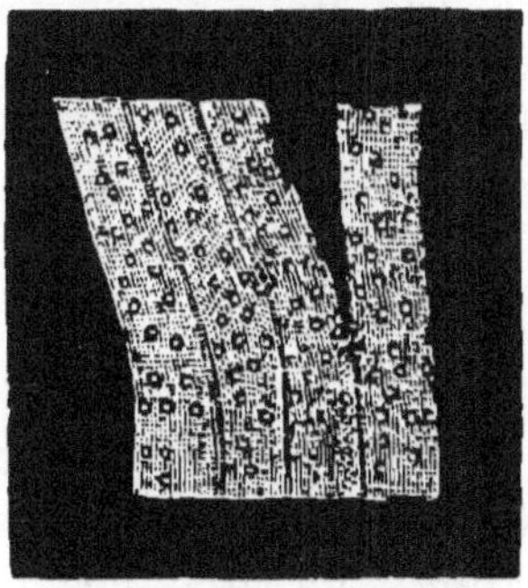

In the second stage (fig. 2) the traces of transversal striæ have almost disappeared; only longitudinal and not undulating fibres are seen; in the primary and secondary bundles (fascicles) the granules are very abundant.

Fig. 3.

The striæ disappear completely; the remaining fascicles, consisting of fibres, only diminished in number, are covered with a still larger quantity of granules (fig. 3); the inter-fascicular spaces are filled

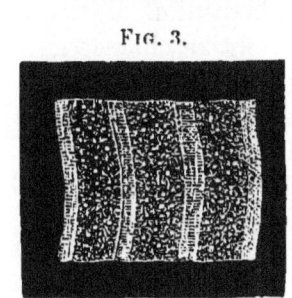

* The first four woodcuts are copied from Laborde.

with the compact fibres of the connective tissue, with a few nuclei dispersed.

In the fourth stage only the skeleton of the muscular bundles is seen; these contain but traces of longitudinal fibres; granules are predominant, and fill the bundles, of which only a very few remain. The inter-fascicular spaces being still larger, contain only the abundant fibres of the connective tissue.

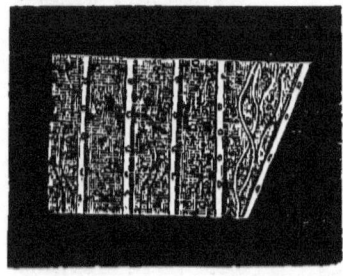

Fig. 4.

In the last stage all traces of muscular tissue, as well as of the granules, disappear; only the empty transparent and hyaline tubes of the myolemma remain with a few granules on their walls (fig. 4). These tubes are separated by large spaces, and are surrounded by fibres of connective, fibrous, and elastic tissue.

The last two stages are found where the muscles had entirely disappeared, and had been replaced by a kind of whitish band.

Although my extracts on the pathological anatomy of this remarkable disease have been very shortened, they demonstrate clearly the changes which take place in the spinal marrow, they prove that there is a material lesion in a central nervous organ, show the progress of atrophy; further, that, notwithstanding Duchenne's authority and diagnosis, no fatty degeneration has taken place, and that the substitution of the fatty tissue for the muscular is not an essential characteristic of the localising atrophic infantile paralysis. The fatty degeneration is another form of the disease, the various stages of which are seen in the following diagrams.

(v) *Fatty degeneration.*

The following diagrams* show the various stages of the

* These have been copied from those drawn and described by Dr. Mandl, and published in Duchenne's *Electricité localisé*, first edition, 1855, pp. 558-559.

fatty degeneration, microscopically examined, and the difference from the *granular* degeneration.

Fig. 5 are the normal fibres, with transversal stripes; traces of longitudinal fibres are still seen; such fibres have been found in muscles which had retained their colour, and during life voluntary and electric contractility, as observed by Duchenne.

Figs. 6 and 7, the transversal stripes are less distinct, and in many places interrupted; here and there they disappear, and finally no trace of them is left; but the longitudinal fibres are more and more visible.

Fig. 8, the muscular bundle is composed only of longitudinal fibres; the transversal stripes have entirely disappeared. Outside the muscular fibres fatty tissue is seen, which consists (*a*) of round or longitudinal cellules. Besides these there are (*b*) small drops (gouttelettes) of fat deposited *within* the muscular fibre.

Fig. 9, the longitudinal fibres, having retained contractility, appear undulated.

Fig. 10, the longitudinal fibres are less distinct; the molecules of fat are more copious; and in fig. 11, these fat molecules being still more numerous, cover almost entirely the muscular fibres.

Fig. 12, the longitudinal fibres have *entirely* disappeared, the fat molecules are very dense and not very distinct, especially towards the axis of the muscular bundle.

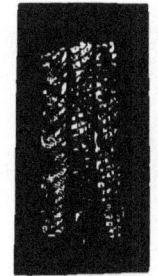

FIG. 5.

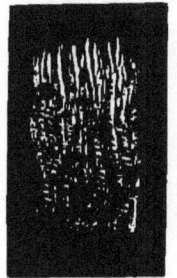

FIG. 6.

FIG. 7.

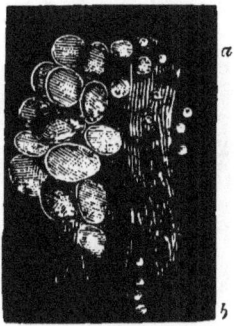

FIG. 8.

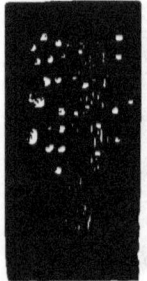

FIG. 9.

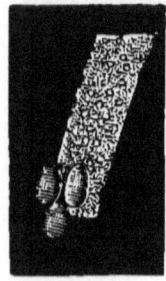

FIG. 10.

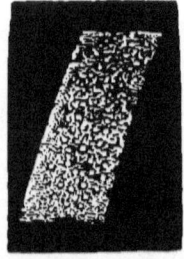

FIG. 11.

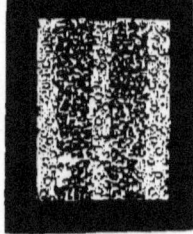

FIG. 12.

Fig. 13, the fat is still more abundant and more diffused, which causes the greater transparence of the muscular bundle.

Fig. 14, the molecules of fat cannot be seen distinctly; the whole bundle consists of an amorphous substance.

Each stage of fatty transformation is marked by an increased discoloration of the muscular fibre, that is, the change of muscular texture is in a direct ratio to its discoloration.

8. *Paralytic deformities.*

The following are the most frequent deformities caused by the atrophic and localising paralysis.

When any of the four principal or of the intermediate abnormal positions of the foot is permanently caused by any pathological process, the foot is called clubfoot, talipes.

The various deformities of the foot are classed according to the direction of the axis of the foot, which can be moved down by extension (*equinus*), inwards by adduction (*varus*), outwards by abduction (*valgus*), and up by flexion (*calcaneus*). Besides these there are the intermediate positions caused by combination of flexion and extension with abduction and adduction; thus *equinus-varus* and *equinus-valgus* are caused by predominant extension of the foot, with the point directed inwards or outwards. If the adduction or abduction is the predominant form, and the extension less developed, the deformity will be called *varus-equinus* or *valgus-equinus*. *Calcaneus-varus* or *calcaneus-valgus* is caused by predominant flexion of the foot, while the point is directed inwards or outwards.

Flexion and extension takes place in the joint formed by the astragalus and the tibia and fibula.

The adduction and abduction takes place in the joint formed by astragalus and calcaneus, and in the medio-tarsal joint, whilst the intermediate movements act on all these joints simultaneously.

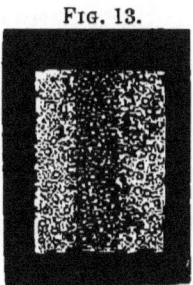

Fig. 13.

Fig. 14.

9. *Talipes varus equinus*

Is the most frequent. The foot is in a state of adduction and extension, the point of the foot is directed inwards and upwards, the external margin of the sole of the foot presses on the ground, while the internal margin of the plantar region affords no support. When no curative means are employed the inversion of the foot increases till the internal margin is directed upwards and forms a right angle with the tibia. The internal ankle is not visible, whilst the external projects, appears lower and more backward. The dorsum of the foot is very convex, whilst the sole is very concave and furrowed lengthways. The weight of the body rests on the rounded, almost half circular external margin of the foot, the skin and cellular tissue covering this part becomes callous, and serves for the sole. The big toe is usually abducted and separated from the other toes, whilst the sole is turned inwards, the dorsum outwards, the inner margin upwards, and the external margin downwards. The constantly tense tendo Achillis is directed obliquely from without *in*wards; the calf of the leg is never formed. The heel drawn up and inwards is the posterior extremity of a curve formed by the big toe and the internal border of the foot; and when the patient stands the heel does not touch the ground.

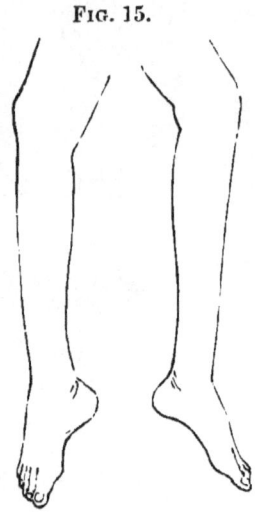

Fig. 15.

Fig. 15 represents paralytic clubfeet. The right foot is varus equinus, of a boy affected with paraplegia and paralysis of the erectors of the spine. The patient was sitting and the legs hanging down when the drawing was made.

Fig. 16 the legs of a boy eight years old; the atrophy of the right leg is very considerable, and the right foot is an equinus varus; the left foot is slightly equinus valgus, and without any atrophy. This patient was not able to bend the right knee.

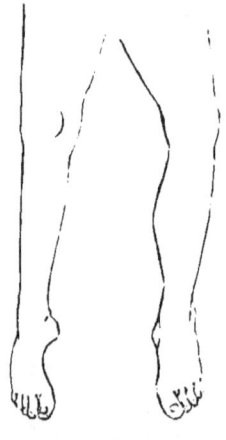

Fig. 16.

10. *Talipes equinus*

Is the second next in frequency, and is very frequently combined with a slight varus.

The characteristics of the pes equinus are that the foot, being in extension, forms almost a perpendicular line with the leg. The heel is drawn up, and the patient walks on the plantar surface of the toes; the sole is very concave, the dorsum convex. In proportion to the longer duration the toes become much stronger than the rest of the foot. The muscles and bones of the leg are imperfectly nourished, the tendo Achillis is very tense, and the muscles of the calf very shortened; in some cases the extensors of the toes are paralysed and the flexors sound. In these cases the patient walks on the back of the toes.

Fig. 17 is a more advanced paralytic equinus with extension of the metatarsal part of the big toe and *flexion* of the last joint.

Fig. 17.

Case.—Miss ——, about eighteen years old, dark, apparently very well built, was paralysed in consequence of her father refusing to allow her to marry some one whom she loved. She had lost the use of the right arm and leg. The extensors of the forearm and fingers having lost their power, the elbow was contracted, and the flexors of the foot having been paralysed, she had a pes equinus, and the gastrocnemii were to such an extent contracted that the mother consulted me about a tumour in the calf. External power or manipulation was scarcely sufficient to replace the foot in its normal position. But I succeeded in proving that there was neither swelling nor tumour, by inducing her to stand with her entire weight on the right leg only, which caused a passive extension of the muscles of the calf, and the disappearance for the time of the spurious tumour; this case was also remarkable from her incapability of speaking, and instead of an affirmative answer to any question she said only "certainly," but she was able to *repeat* words and small sentences which I said first. She understood all she was told, but it appeared that her power of expressing her ideas by words was lost.

11. *Talipes valgus*

Is less frequent: the foot is in abduction, the toes are directed outwards, the internal ankle projects, the internal margin of the foot with the big toe presses on the ground, whilst the external border of the foot gives no support. When the complaint is not arrested the deformity of the foot increases to such an extent that the patient walks on the malleolus internus, and on the inner part of the dorsum of the foot. The cellular tissue and the skin covering these parts become callous; this deformity is also combined with an inversion or incurvation of the knees (genu valgum), of which I have had several cases under treatment.

Fig. 18 is a more advanced stage of valgus; the eversion of the foot is very considerable.

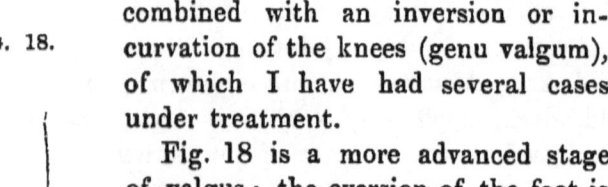

FIG. 18.

12. *Talipes calcaneus.*

This form is rare: the foot is in flexion, the dorsum forms with the leg a more or less acute angle. The gastrocnemii are relaxed; the weight of the body rests entirely on the heel, which is the only part of the foot which touches the ground. The anterior part of the foot is raised; when the extensors of the toes are paralysed, a kind of arch is formed in the sole of the foot, and this is one of the combinations of this form of clubfoot. But it is very rarely that the calcaneus is not associated with a slight abduction or adduction by which the calcaneus valgus and varus is formed.*

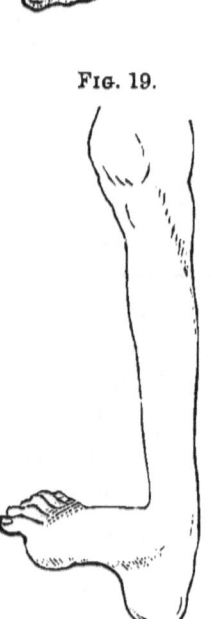

FIG. 19.

Fig. 19.—Paralysis calcaneus in a more advanced stage, with contraction of the planta pedis.

---

* According to the predominant character of the various forms of clubfoot, the deformity is named. Thus, equinus-varus means that varus is only a combination or an addition to the equinus, whilst varus-equinus specifies that varus is predominant and equinus accessory. There is even a variety of valgus-varus or varus-valgus; in these cases the first word designates the character of the deformity in the ankle-joint, whilst the position of the toes is in a contrary direction.

13. Congenital clubfoot.

There are artificial classifications of the varieties of clubfoot. For practical purposes the forms I have mentioned are sufficient.

I have only to add that clubfoot is also *congenital* and acquired, and traumatic. In the congenital form there are frequently also deformities of the bones of the foot present; if only the muscles are unequal in their action, atrophy to a very considerable degree is rare. After protracted neuralgia, rheumatism, burns, caries of the bones, ulceration of the soft parts of the foot, and especially in strumous constitutions, clubfoot is not rare.

After violent external injuries, clubfoot has sometimes the character of paralytic clubfoot.

FIG. 20.

*Pes valgus congenitalis.**—Fig. 20.

(*a*) The deviation of the axis of the foot. The heel and the external margin of the foot are raised.

(*b*) A projection caused by the malleolus internus, astragalus, and os scaphoideum.

(*c*) Creases of the skin which hide the tendo Achillis, and underneath which the heel is deviating considerably outwards.

(*d*) Malleolus externus is scarcely visible.

Pes equinus congenitalis.—Fig. 21. (Delpech.) Slightly complicated with *varus*.

(*a*) The deviation of the axis of the foot towards the plantar surface and the internal margin.

(*b*) The toes on which the whole weight of the body rests.

---

* *Delpech Chirurgie Clinique de Montpelier*, tome i, Paris and Montpelier, 1853.

(c) A projection formed by the malleolus externus and astragalus.

(d) A prominence formed by the cuboid and cuneiform bones, and the posterior (but here upper) ends of the metatarsal bones.

(e) Creases of the skin covering the very tense tendo Achillis.

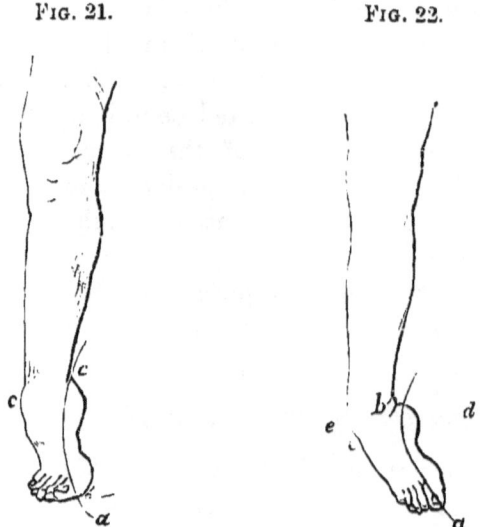

Fig. 21.   Fig. 22.

Varus equinus congenitalis.—Fig. 22. (Delpech.)

(a) Shows the torsion of the foot, deviation of the axis of the foot (down inwards and towards the external margin).

(b) Creases of the skin covering the tendo Achillis, which is deviating from the normal direction.

(c) That part of the back of the instep and of the toes on which the weight of the body rests while walking.

(d) Malleolus internus slightly developed and hidden in the tibio-tarsal articulation.

(e) Malleolus externus very prominent.

Fig. 23.—Congenital left pes equinus varus in a child of six months.

Fig. 24.—Right foot of the same child.

FIG. 23.   FIG. 24.

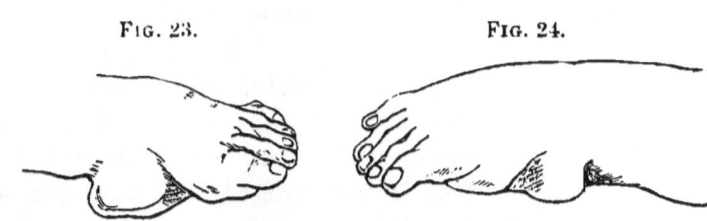

Fig. 25.—More advanced state of congenital pes varus; both legs are drawn in their natural position up to the thighs, in order to show the disproportion in the knee-joints, as well as the relative position of both feet towards each other; the feet are placed with their toes almost opposite to each other, and are remarkably broad and short, especially in the region of the heel, and in the middle of the instep. The arch of the instep is large; and the rotation along the *outer* margin considerable; the knee-joints are turned inwards; the condyli femorum interni are opposite each other; the patellæ do not participate in the rotation, but are placed as usually on the anterior surface. Thus the knee-joints appear to be dislocated. The small size of the patellæ is remarkable, which fact was first observed in those suffering from clubfeet by Ammon.

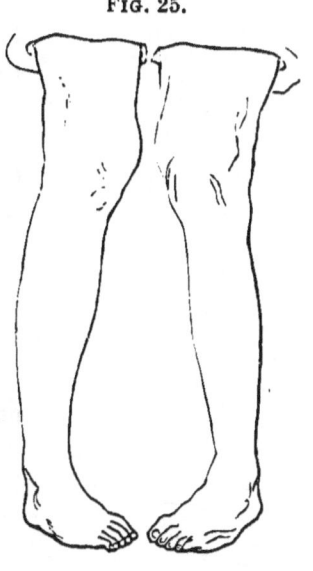

FIG. 25.

The diminished development of the knee-pan is explained by the patient, when he walks, using his legs by raising them in the hip-joints, whilst the knees remain more or less stiff; and the feet moved in a segment of a circle are placed alternately one before the other.

Talipes calcaneus congenitalis.

Fig. 26.—The foot is bent upwards to such an extent that

Pes varus Congenitalis.

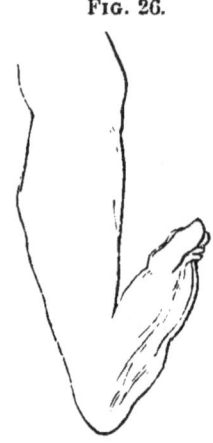

Fig. 26.

Fig. 27.

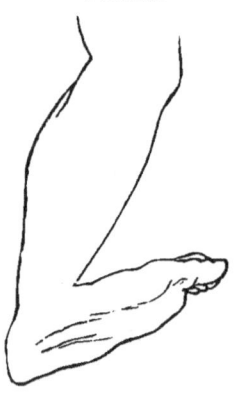

Fig. 28.

the instep almost touches the front of the leg, and the weight of the body is supported when walking by the posterior tuberosity of the calcaneus. The toes are bent, but the foot is not twisted to either side, as in valgus or varus.

Fig. 27 also represents a congenital calcaneus, but the acute angle formed by the foot and leg is smaller than in fig. 26.

Fig. 28 is the posterior aspect of a talipes calcaneus, with the elongated tendo Achillis. (Ammon.)

Pes varus congenitalis.

Fig. 29 represents a more advanced degree of right club-foot. The foot is broad, short; the malleolus externus very much projecting; the muscles of the calf are very tense.

Fig. 30 is a similar left club-foot. The development of the external margin of the foot which serves for the sole is more marked.

Figs. 31 and 32 represent the two feet (figs. 29 and 30) seen from the sole.

The soles are remarkably broad, and that part of the foot nearest to the toes is very strongly developed; the great tension of the tendo Achillis is more seen.

Pes varus Congenitalis.

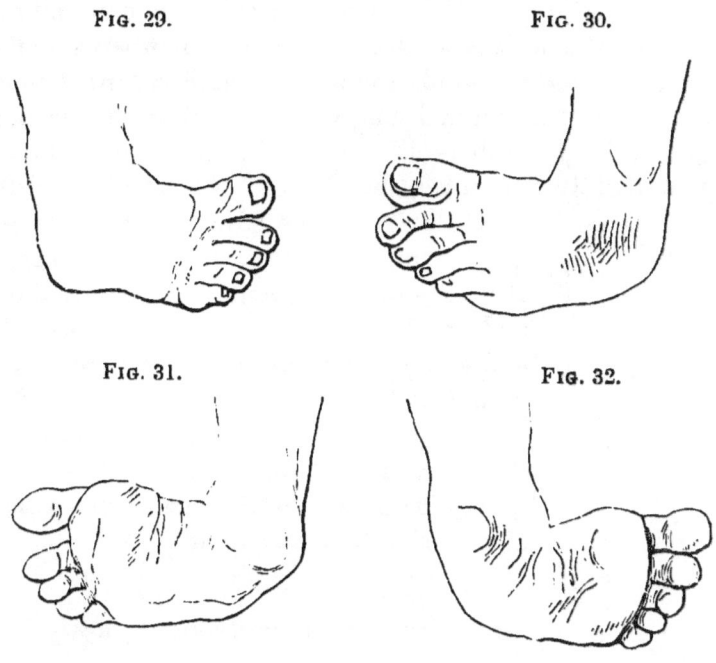

FIG. 29.   FIG. 30.

FIG. 31.   FIG. 32.

Pes valgus congenitalis.

FIG. 33.

Fig. 33.—A fœtus of six months. The dorsum of the foot is not vaulted, but flat and broader than usual; the internal margin has not its natural curve; the sole touches the ground in its entire surface; the heel is prominent and directed slightly inwards.*

The paralytic deformities of the knee-joint.

14. *Genu incurvatum* (valgum) *paralyticum.*

The knee forms a slight arch, with the convexity inwards towards the mesial line of the body. The thigh-bone is placed in an obtuse angle to the leg, which *slants* outwards.

* All figs. of congenital deformities are copied from Dr. Ammon's work, *The Congenital Surgical Diseases of Man*, Berlin, 1842.

If both legs are simultaneously affected, the deformity is usually known as *knock knees*. The patient when standing presses one knee towards the other; the feet cannot touch each other at the internal margin; and with more advanced stages a talipes valgus is developed; the hip-joints are usually slightly contracted, and this causes what is called the *cock's gait*; in the advanced stage the adduction of the knees attains such a degree that it is scarcely possible to separate them, and in this case the patient cannot walk, and shuffles himself along with difficulty, even when he has sticks or crutches. The abductors, the rotatores, and the glutei are the muscles most frequently paralysed.

Fig. 34.

Fig. 34 is a paralytic genu incurvatum combined with pes equinus.

15. *Genu excurvatum* (varum) *paralyticum*

is a sort of bow-leg, but not from rickets (although the combination of rachitis and paralysis occurs sometimes); there is no deformity of the long bones of the lower extremity, but a slight arch is formed in the knee-joint, with the convexity *out*wards.

The adductors of the thigh are more or less paralysed; the ligaments of the knee and some joints extremely relaxed; if both legs are simultaneously affected, the knees are contracted by retraction of the flexors of the leg, the vastus being also paralysed; the legs and knees are, as soon as they are placed together, separate, and fall again outwards. Fig. 40 shows this state, which the French call the Polichinelle's legs.

16. *Genu recurvatum*

is a less frequent deformity; in the normal state, when the foot is placed straight in front of the tibia, and the leg well stretched, a slight arch is formed in the knee,

Paralytic Deformities of the Knee-joint.

with the convexity *back*wards; this arch is considerably increased when the flexor muscles of the leg are paralysed, and the ligaments in the knee-joint relaxed; this abnormal convexity, of which the knee is the highest point, forms the characteristics of this deformity.

Fig. 35.

Fig. 35 is a paralytic genu recurvatum with pes equinus.

17. *Contracted knee.*

The angle formed by the leg and thigh varies from an obtuse to an acute angle.

The extensors of the leg are paralysed.

This deformity rarely occurs alone, but in combination with club-foot, contracted hip, &c.

The paralytic deformities of the knee-joint vary also often in both legs, and one genu is (valgum) *in*curvated, the other (varum) *ex*curvated. Sometimes one knee is only affected, whilst the other leg is either entirely sound, or a club-foot present.

Fig. 36.

Fig. 36 is a paraplegia, with contracted hips and knees; the right is genu excurvatum, the left genu incurvatum.

18. *Paralytic affections of the hip.*

Contracted hip.

When the glutei are paralysed, the psoas and iliacus internus being without antagonists, cause the deformity

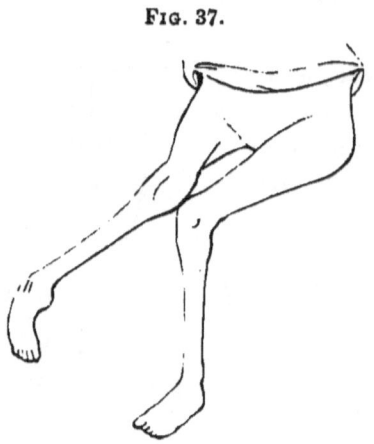

Fig. 37.

which is usually combined with an inward turned position of the thigh, in consequence of the muscles which turn the thigh outwards being weakened; if both hips are affected incurvation (lordosis) of the lumbar part of the spine is often present.

Fig. 37 is a case of paraplegia with contracted hips, knock-knees, and varus of both feet; the atrophy of both legs is very far advanced.

19. *The relaxed or loose hip.*

When all the muscles round the hip-joint are paralysed, and the ligaments relaxed, the whole leg dangles and swings about like a lifeless part; turning or bending of the lumbar part of the spine is the only mode by which the hip is apparently moved, or rather *pulled* in the direction of the spinal movement. The thigh can be moved passively in all directions without causing the patient any pain, the outlines of the hip-bone and femur (the hip-joint being covered only with the atrophied glutæi) are easily visible or felt under the skin. It is one of the most serious and tiresome processes to improve this state of the hip-joint, especially when it occurs simultaneously in both hip-joints and when combined with a paralysis of the whole lower limb.

Paralytic affections of the spine.

20. *The incurvation of the spine, or paralytic lordosis.*

This takes place mostly in the lumbar part of the spine; the erectors of the spine are paralysed, while the psoas remains active. This deformity occurs frequently with paraplegia; when the patients have not lost all power of stretching their knees, they try to walk while bending the body almost in a right angle with the thighs and while the arms, usually very strong, are placed on chairs or on two sticks. In fact, they assume the position of a quadruped whilst walking, and during this action the indenture of the spine is particularly strong, while the hips seem to rise and fall alternately with each step they are taking in a *lateral* direction.

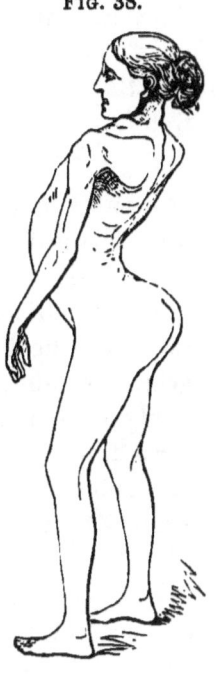

FIG. 38.

Fig. 38 represents a paralytic lordosis; the patient is not able to stand except when supported and prevented from bending or rather falling forwards.

21. *The lateral curvature of the spine*

is observed in hemiplegia, in paralysis of one extremity, and especially of those joints by which the limb is attached to the trunk, and of the muscles of one side of the trunk which are more or less atrophied.* The shoulder-blade and the hip-bone are sometimes arrested in their growth, and the careful observer will find a great difference between the affected and the healthy part. When a leg is shorter, or the power of standing on one leg, of stretching one knee, and of using one ankle-joint is diminished or in any way interfered with, lateral curvature must be formed as a compensation for the deficient functions. The majority of these compensating forms of lateral curvature are easily prevented by supplying the deficient length of the leg, by supporting contrivances which prevent the deviation in the knee and ankle-joint.

* Mr. Barwell attributes the majority of lateral curvatures to paralysis of the serratus major on one side. Neither my own experience nor the physiological study of the functions of this muscle by Duchenne confirms this hypothesis.

22. *The paralytic kyphotic curvature.*

This occurs only in those cases where the little patients are to such an extent paralysed in the limbs as well as in the trunk that they are unable to sit. The spine forms one round arch from the neck to the os sacrum with the convexity backwards; *kyphotic* means here only this direction of the spine, but not an angular projection as it occurs in Pott's disease.

23. *The paralytic deformity of the neck.*

Amongst children cases occur in which the head falls down, forwards, and slightly sideways, in consequence of the paralysis of some of the muscles of the neck; paralysis of one sterno-cleido mastoideus causes wry necks without disease of the ligaments, intervertebral substances or the cervical vertebræ. Cases of *paralytic* wry neck due to neuralgia or rheumatism, have been observed by myself several times in children and adults; very considerable single and double lateral curvatures have been caused in some instances through neglect of attending to the primary wry neck.

The deformity of the shoulder by atrophy of the deltoideus has been already mentioned; that which is caused by atrophy of the large pectoralis I have not seen. The paralysis and atrophy of the muscles which adduct the shoulder to the spine, or pull it downwards, causes that particular position of the shoulder-blade where the internal edge of this bone is directed upwards, and the lower angle projects backwards above the latissimus dorsi. In Duchenne's *Electricité localisé* several cases of this deformity are described.

24. *Paralytic deformities of the elbow and wrist-joint.*

Contraction of the elbow-joint, with inversion of the forearm and contraction of the wrist and fingers, is the most frequent form in cerebral paralysis, especially in hemiplegia. In a case of atrophic localising paralysis which came under treatment eighteen months after the invasion of the disease, even the passive pronation and supination of the forearm was entirely impossible; in the same arm there was eversion

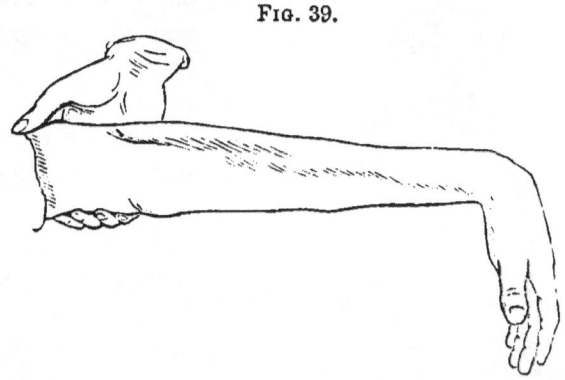

Fig. 39.

of the wrist; the dorsum of the hand formed a right angle with the forearm; while the paralysed fingers, in appearance stout and fat, were constantly bent. There are other forms in which the hand is deformed analagous to the various forms of club-foot; the fingers are also deformed in one or several joints, mostly contracted, sometimes in the form of bird's claws, according to the various muscles of these parts which are paralysed; sometimes the ligaments are relaxed to such a degree that the fingers can be bent towards the dorsal surface as easily as towards the palmar side. Fig. 39 shows the paralysis of the extensors of the wrist in a boy who had cerebral hemiplegia, afterwards epileptic fits. The deformity of the wrist is not yet removed.

25. *General paralysis.*

Fig. 40 represents a boy whose limbs are still paralysed, after having recovered his power of moving the neck and back. It is one of the cases in which the adductors of the thighs are powerless, the knees slightly contracted, the feet are talipedes equini. This child's legs can be placed together with the knees touching each other, but they must be bound or held together, otherwise they return immediately into the position shown in the engraving. The atrophy of the deltoideus on the left side is shown by the greater projection and the more distinct outlines of the left acromion.

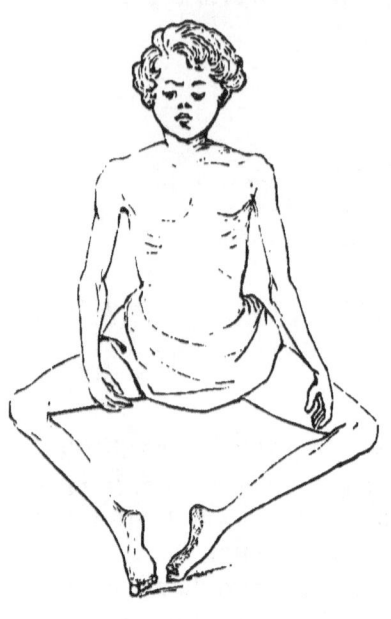

Fig. 40.

The outlines of the legs and of the arm show the more advanced progress of the atrophy of these parts.

The few engravings will be sufficient to give at least a general idea of deformities caused by paralysis. I could not enter into a minute and more graphic description, and have merely pointed out the most frequent forms which very rarely affect one joint only, or a single part only. The combinations are very numerous and I may state, as a rule, that the single deformities increase not only in intensity, but that their combination with others will be more numerous in proportion to the length of time the suitable means for their treatment are neglected.

Much precious time is lost by a still prevailing fancy even amongst those who are considered to be in the foremost ranks of the profession, that the child will grow out of its deformity, and again by the indiscriminate application of rubbing, electricity, mechanical contrivance, and strychnine. If the rubber styles himself a medical rubber, eminent medical men will have their patients rubbed without giving any special instructions what kind of friction, manipulation, or movement should be done in the individual case.

How electricity and orthopædic contrivances are abused will be mentioned under the head of these agents. In consequence of the good results obtained by *Nux vomica* and *Strychnine* in some forms of peripheric paralysis, these drugs are constantly prescribed in paralytic affections from central causes, although the majority of medical writers agree that they have not answered their expectations.

26. *Diagnosis.*

The (atrophic and localising) infantile paralysis
which is the most frequent source of shrivelled, deformed, half-dead limbs, clubfeet and clubhands incurved, excurved, recurved, and contracted knees, and of many other sad deformities, has the following characteristics:

1. In many cases a feverish state of longer or shorter duration precedes; sometimes gastric typhoid remittent fever, measles, scarlatina, whooping cough, pneumonia. In other cases no fever (or of such a slight character that it is not noticed) is observed.

2. The access of a more or less general paralysis is sudden, and whilst the infants are in good health, and its intensity is not in proportion to the feverish state or to the slight convulsion if either of these two conditions precede it.

3. A progressive tendency of the general paralysis to localise itself in one or two limbs, or of the paralysis of a whole extremity, to localise itself in some groups of muscles or even in a single muscle.

4. The prevalent flaccidity and flabbiness of the paralysed parts.

5. The progressing atrophy of the muscles, ligaments, and bones.

6. The diminution of temperature.

7. A more or less prevalent deep blue-coloured cyanosis in the paralysed parts.

8. Tendency to chilblains in the affected limbs.

9. Deformities of various kinds of which the most frequent are clubfoot, especially—

(*a*) Talipes equinus; talipes varus; talipes valgus; talipes calcaneus; talipes equino varus.

(*b*) Genu varum with talipes equinus, and if this deformity affects both legs, it gives the appearance of cardboard figures with moveable limbs when the string pulls the thighs outwards and the legs are dangling.

(c) The deformity produced by the atrophy of the deltoidcus which gives to the shoulder-joint an angular or flat appearance.

10. Electro-mobility in the beginning very slight, later entirely wanting; no movement by reflex action.

11. Complete loss of any influence of the stimulus of the will on the paralysed parts, and consequent impossibility of any voluntary movement in the paralysed muscles.

12. Sensation normal; sometimes hyperæsthesia in the early stage proportionate to the subsequent loss of power. I have observed local hyperæsthesia several times near the principal trunk of the nerves of the paralysed extremity.

13. Actual or comparative arrest of development in the paralysed parts.

14. Fatty degeneration not a necessary consequence of this kind of paralysis.

15. Want of rigidity near the deformed joints in the first stages.

16. Sphincters not (or only exceptionally) affected.

17. Usually strong and well developed children; the trunk and upper limbs in comparison to the pelvis and lower extremities much developed, except in cases where an arm is paralysed.

18. Talipes varus of one foot, and talipes valgus often a pathognomonic symptom of this paralysis. (Laborde.)

19. The paralysis scarcely ever takes the form of hemiplegia, and even then does not affect the face; when it does, it is always preceded by paralysis of all the limbs.

20. The intellectual faculties and—

21. The general health are unimpaired, notwithstanding the persistence of the atrophic localised paralysis.

22. The arms recover their power quicker than the legs.

23. The muscles which waste most rapidly are the tibialis anticus and the deltoideus.

24. After tenotomy, the ends of the divided tendons do not leave a large interval between each other, because there is no active tonic or spasmodic contraction of the muscles as in cerebral paralysis.

27. Diagnostic symptoms in cerebral infantile paralysis.

1. Symptoms of cerebral affections produced by any morbid cause precede, during which spasmodic contraction of one leg and arm of the same side frequently takes place.

2. Several and persistent convulsions, attacks of frequent headache, giddiness, indistinct speech, nervous sensations in the ears and eyes, precede the cerebral paralysis.

3. Hemiplegia is frequent; the little patients look like the image of old people with an arm, and hand, and fingers contracted, whilst the forearm is turned inwards a leg is dragged behind them; the face is also frequently affected.

4. There is a hereditary disposition; the father or mother being affected with insanity, paralysis, epilepsy, dipsomania, &c.

5. Children are constitutionally weak.

6. The shape of the head is irregular, frequently an appearance of stupidity, idiotcy, with strabismus and indistinct speech.

7. The intellectual and sensorial faculties are disturbed.

8. The contracted muscles are rigid.

9. Electro-contractility is normal, and even greater than in healthy parts.

10. No atrophy in the beginning; fatty degeneration and atrophy are sometimes consecutive.

11. There is reflex movement.

12. The temperature of the paralysed parts is diminished, but never so low as in the localising atrophic form.

13. Legs recover their power more quickly than the arms.

14. In the paraplegic form the contraction and deformities of the lower extremities is more prevalent, but notwithstanding the severity of these contractions, the patients are able (when assisted) still to move and even walk in some although very tedious manner, and thus retain a power which is entirely lost in the localising atrophic paralysis.

15. The contracted elbow, wrist, hip, knee, or ankle-joint, *can* be stretched, but only by degrees, and with a great amount of force, sufficient to overcome the rigidity of the

contracting muscles; as soon as the external power ceases the muscles contract again.

16. After tenotomy the ends of the divided tendons have a large interval in consequence of the spasmodic contractions.

17. Sphincters are either partially or completely paralysed, and the water is frequently passed when the little patients scream or are frightened.

18. Involuntary tremor or nervous twitching of the limb, and involuntary movement of the fingers and toes, which are usually spread, are observed.

19. The influence of the stimulus of will is not entirely lost, and a few, although very slow, movements are seen in the fingers and toes, wrist, or instep.

20. The deformities in the limbs are combined with spasmodic contractions, rigidity, and stiffness, and sometimes arms and legs are stretched, the extensors of the forearms and legs being spasmodically contracted.

21. With the improvement of the constitutional health and the disappearance of the cerebral symptoms the deformities improve.

22. Pathological anatomy shows the changes caused by hydrocephalus, inflammation, tumours, &c., in the brain.

28. *Facial paralysis, or paralysis of portio dura of the seventh nerve (this also combined with cerebral hemiplegia).*

1. Takes place suddenly after exposure to cold or without cause.

2. After disease of the temporal bone.

3. In newborn infants after injuries by forceps or a difficult passage at the birth.

4. Eye cannot be closed, eyelids do not move.

5. Escape of tears.

6. Mouth drawn up on the opposite side.

7. Escape of saliva.

8. Labials cannot be pronounced.

9. Cheeks on the affected side do not resist to the blowing out of the walls.

10. Patients cannot whistle.

29. *Diagnosis of paralysis from spinal congestion.*

Dr. Radcliffe* asserts that the (atrophic localising) infantile paralysis is analogous or identical to paralysis from spinal congestion, although only the following symptoms are common to both, whilst they differ considerably in other symptoms.
1. The paralysis is partial.
2. There is hyperæsthesia.
3. Muscles limber, flabby (not rigid).
4. More or less complete recovery is the rule. Negative common symptoms.
5. Sphincters are not affected.
6. Absence of head symptoms.

30. *The difference of symptoms in spinal congestion.*
1. Tingling of the tips of fingers and toes.
2. A dull burning aching in the back and limbs.
3. Being tired to death.
4. A hot sponge increases the pain in the back.
5. Respiration shallow and curiously interrupted by sighs.
6. Tendency to myelitis (to increase).
7. No atrophy.
8. The paralysis is always a paraplegia.
9. Pathological changes: engorgement of the veins of the spinal cord and membranes, with some excess of the spinal fluid; infiltration, with blood of the cellular tissue exterior to the dura mater.
10. The disease has been fatal and quickly fatal in some cases.

The first six symptoms are absent in infantile paralysis; with regard to the seventh, eighth, and ninth, atrophy is characteristic in infantile paralysis, which has also other forms besides the paraplegic, and other pathological changes take place, as diminution of the nervous tubuli, &c. (See Laborde's autopsies.)

* Reynolds's *System of Medicine*, vol 2.

I do not understand why Dr. Radcliffe should have chosen of the ten symptoms of spinal congestion four common to this complaint and to infantile paralysis (because two are negative; viz. absence of head symptoms, and absence of paralysis of the sphincters) to prove the identity of the two complaints; and not to have taken in account the other ten symptoms of which the first six are absent, and three others are different in the atrophic localising infantile paralysis.

31. *Progressive atrophic muscular paralysis.*

SYNONYMS.—Paralysis atrophica, wasting palsy; Cruveilhier's atrophy, *progressive muscular paralysis*.

1. Attacks less frequently very little children, but more frequently young adults.
2. Often caused by cold, exposure to inclement weather, sleeping in damp beds or wearing damp clothes; by some complaints of the spinal cord and by traumatic influences.
3. The invasion of paralysis is always gradual.
4. The atrophy precedes the paralysis, and causes—
5. A certain amount of weakness which increases to perfect loss of power in proportion to the progress of the atrophy.
6. Atrophy more frequent in the upper limbs, the ball of the thumb and hand or the shoulder, are often the starting-points, or the neck, the face, the tongue, the thigh, leg, or foot.
7. One or a few groups of muscles only in an upper or lower extremity are attacked, sometimes several limbs, or all the voluntary muscles of the body attacked.
8. According to the number and position of the atrophied muscles the outlines of the limbs or body are changed, round parts are flattened, bones project.
9. Quivering in the affected muscles diminishes with the increase of atrophy.
10. Electro-contractility continues as long as the atrophy is not too far advanced.
11. There is no loss of temperature.
12. If the muscles of the face waste, the features are not

changed by any emotion, and the patient looks stolid and solemn. If the neck is affected the head falls forward with the chin on the sternum or laterally with head on the shoulder. When the forearm and hand are attacked, the claw-shaped hand (*main en griffe*) is the result.

32. *Diagnosis of myelitic paraplegia.*

1. The paralysis is usually complete and preceded by great restlessness.
2. The sphincters are paralysed.
3. Anæsthesia.
4. No reflex action.
5. No electro contractibility.
6. Absence of pain, tenderness in the spine.
7. Hot sponge or ice passed down the spine causes a burning sensation in the region corresponding to the inflamed part of the cord.
8. The urine is alkaline.
9. The muscles of the paralysed parts are limber.
10. Tendency to increase.

33. *Paralysis after spinal and cerebral hæmorrhage*

is also sudden, with pain in the spine and violent tetanic convulsions in the first, and with loss of consciousness in the second.

34. *Duchenne's pseudo-hypertrophic infantile paralysis, or infantile paralysis with apparent muscular hypertrophy.*

This is a form of paralysis to which Duchenne was the first to call attention, and is a form of progressive muscular paralysis, with enlargement of some muscles, especially of the gastrocnemii, which appear to be hypertrophied, although the increased size is only due to a deposit of fat and areolar tissue between the muscular fibres. About thirty to forty cases have hitherto been described in France, England, and Germany. Two cases were brought before the Pathological Society by Dr. Hillier and Mr. Adams. Duchenne published last year a little book, *La Paralysis Pseudo-hyper-*

trophique. He impresses upon the profession the importance of recognising *this disease at an early stage*, and has invented a little instrument to pull out a few muscular fibres from the living patient in order to submit them to microscopical investigation. The disease being progressive and fatal, he asserts that electricity can arrest its development, and thus save the patient. He gives the history of a patient whom he cured. The diagram (fig. 41) which illustrates this form of paralysis is taken from his book, and will convey better than any description the characteristics of this complaint, of which the following are the principal symptoms.

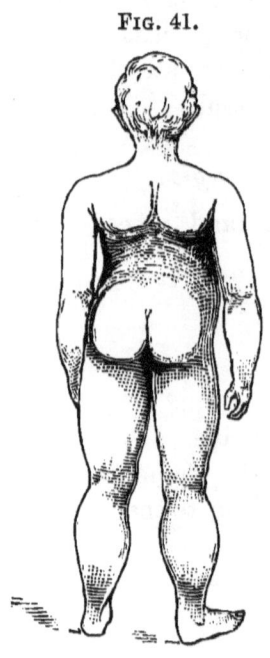

Fig. 41.

35. *Pseudo-hypertrophic paralysis.*

1. It occurs always in children, and is noticed when they ought to begin to walk.

Those who have been already able to walk feel an increasing difficulty in walking.

2. There is a tendency to arch the lower dorsal and lumbar spine forwards, that is, to form a lordosis of these parts.

3. When an attempt to walk is made the legs must be separated to preserve balance.

4. The power of moving the legs decreases slowly.

5. The upper extremities are thin and get weak.

6. The respiratory muscles lose power.

7. The muscles of the lower extremities, especially gastrocnemii and glutei enlarge, as well as the muscles of the back; they are elastic, but very firm, which state may last two years.

Duchenne's Pseudo-hypertrophic Paralysis.

8. This enlargement is caused by a deposit of fat and areolar tissue between the muscular fibres.

9. The muscular tissue of the enlarged parts wastes, the interstitial deposit is absorbed and atrophy results.

10. Involuntary muscles and sphincters are not affected.

11. Hitherto no disease of the nervous system has been observed.

12. It is probable that the cause of the abnormal nutrition of the enlarged parts is caused either by some change in the spinal cord, in the sympathetic nerve or in the blood.

The following is an outline of Hillier's case which I have before named. The boy was eleven years old, his parents in good health; although he walked when twenty-one months old, he could never walk well, neither run nor jump; he waddles with feet apart, and tumbles when trying to walk quickly. His calves have always been large and out of proportion to the other muscles.

Since his third year the power of walking has declined; during the last six months he has been quite unable to walk, he cannot raise himself up in bed, nor stand unless supported; his heels are drawn up, the lower part of the spine is arched forwards when trying to stand. Lately the upper very thin limbs are failing; he cannot cut his food or grasp anything firmly: there is complete control of the sphincters.

Intellect neither very bright, nor very deficient; slow to learn. The calves are still a fair size, but have a peculiar doughy feeling.

Dr. Foster, on February 26th, 1869, in the pathological and clinical section of the Birmingham and Midland Counties Branch, presented a boy aged 9, an excellent example of Duchenne's paralysis, with apparent muscular hypertrophy.

The boy was in the second stage of the disease, the glutei and gastrocnemii enlarged, firm and bulging in contrast to the poorly developed muscles of the upper extremities. The spinal extensors were hypertrophied, and in the erect position the characteristic lumbo-sacral curve was well marked. The boy waddled rather than walked, balancing himself on each leg alternately. When placed on his back he raised himself by turning over, obtaining a point of support for his feet, and then lifting himself by placing his hands on his knees and thighs.*

36. Traumatic paralysis.

Loss of power; dropping of the injured part of hands, feet, or a whole limb. Atrophy of these parts, sometimes followed by fatty or granular degeneration; lower temperature; pain, swelling, congestion, inflammation and erysipelas soon after the injury; sometimes loss of sensation of cold, heat, contact, or morbid sensations; glossy fingers or toes; arrest of development in growing persons; various deformities similar to those in other paralytic affections dependent upon the various parts which have been impaired.

In traumatic paralysis, as in a local infantile paralysis, the paralytic symptoms mark the outset of the disease, then after a shorter or longer period the muscles supplied from those parts of the cord least affected recover their voluntary movement and nutrition, whilst those which receive their nervous supply from parts more profoundly injured are atrophied, and sometimes fatty or granular degeneration follows.

A little boy nine or ten years old was violently pulled by the arm when an infant by his little brother, and from that period for a long time after the limb appeared powerless. He gradually recovered some use of it, and could move it in any way, and was fond of climbing with it. It was, however, comparatively very weak, and was small like one extremely emaciated. It was not short, but altogether not more than two thirds the size of the other. (Paget.)

A ladder fell on the back of the shoulder of a little girl seven years old, and then broke her leg. The exact manner in which the ladder fell was not known, but the integuments over the scapula and by the side of the neck were severely bruised. She was stunned and unconscious

---

* A further account of this case and of its treatment, which had not been successful, is published in the *Lancet*, May 7th, 1869.

for less than ten minutes, and then perfectly recovered her senses. As soon as she recovered from the shock she called out " Where's my arm ?" and from that moment to the time when first seen by Mr. Paget, about four months after the accident, there was perfect insensibility of that limb.

It had been for a time painful subjectively, and there were some kinds of contact which distressed her, but she could feel no common touch, no heat nor cold, and had only morbid sensations either spontaneously or from some irritation. There had been also total loss of motion in the arm until within three months after the accident, when the pectoral and posterior scapular muscles had regained slight power, every part of the arm had greatly wasted, and it was habitually cold, with slight swelling and congestion of the hand.

37. *Arthritic and rheumatic paralysis*

is not so rare amongst the young as is generally believed.

Electro-contractibility continues; pain in the muscles, ligaments, or joints. The constitutional symptoms are the most essential characteristics.

38. *Diphtheritic paralysis.*

The preceding diphtheria, the affection of the fauces, uvula, velum palati, impossibility of swallowing, and the greatest weakness, characterise this complaint, which improves in proportion to the recovery of the general strength, although the rate of mortality is very great among children affected with this form of paralysis.

39. *Rickety paralysis.*

Besides the constitutional signs, the enlargement and deformities of the bones, of the joints, of the spine and ribs, the large head and the expression of premature age are the principal symptoms, the paralysis is incomplete, and there is a gradual increase of muscular weakness; other forms of paralysis are combined with rickets in cachectic children.

40. *Paralysis through Potts' disease.*

Kyphosis of some diseased part of the spine, pain in the spine, psoas, lumbar, and other abscesses, usually paraplegia with rigidity of the legs, paralysis of the bladder, sphincter ani, loss of motion and sensation in the legs, swelling of the scrotum, constipation, are the most common symptoms—many of these as well as the paralysis disappear with the progress of recovery, and when the pressure on the spinal cord or its nerves caused by exsudation or suppuration of the surrounding part ceases.

41. *Paralysis after chorea.*

Protracted chorea frequently causes paralysis of those parts which have been involuntarily moved and jerked about. It may be that chorea is merely concomitant, or associated with paralysis and not its cause. Hemiplegia with chorea and slight impediments of the speech is the form which I have observed several times; unsteadiness in walking, shaking the head, bending the body forwards to such an extent that the lumbar concavity of the spine is effaced, while one or both arms with the hand twisted inwards are kept backwards to balance the body; incapability of stretching the knee of one leg, an involuntary or irresistible tendency of walking very quickly, almost of running, which causes sometimes very dangerous falls, considerably diminished power of using one side of the body with the corresponding limbs, deficient power of balancing, are some of the characteristics of this complaint, to which must be added a heavy tongue, a brawling speech, stuttering and stammering in some words, coldness of the extremities and diminution in the size of the weak limbs, although this does not amount to atrophy.

42. *Paralysis from epilepsy and convulsions.*

Hemiplegia, paraplegia, and general paralysis are the most common forms. In the hemiplegic form, the deformity of the arm and hand is caused by the paralysis of the extensors; the elbow, wrist, and fingers are mostly contracted, and the fore-arm inverted, the hip and knee contracted, the latter turned inwards, with pes equinus and slight varus.

In the paraplegic form the hips are contracted, adduction of the knee causes the angular incurvation of the knee, and valgus of both feet. When the patient is able to walk he throws himself from one side to the other, and one knee is placed before the other. The character of the gait is *waddling*. If this complaint is left to itself the knees can be scarcely removed from each other even by external force.

Of general paralysis after convulsions, the following is one of the forms I have observed. The whole body appears more or less rigid, the flexors of the elbow- and knee-joints on both sides are paralysed, and all the limbs are stretched and stiff; the thighs are in a straight line with the body; pes equinus varus of both feet; both hands contracted in the wrists. The degree of paralysis and deformity is not equal on both sides; the intellectual and perceptive faculties appear slightly impaired, but the power of speaking is more or less impaired.

43. *Paralysis produced by metallic cosmetics.*

Pearl powder is used by nurses and vain mothers to improve the skin of little girls; adult girls use this powder, as well as other cosmetics, into the composition of which preparations of lead and mercury enter; weakly constituted girls are liable to absorb the poison which, by degrees, produces its effects.

Carbonate of lead has been, and is, sold as pearl powder. Dr. Cousins, of Portsmouth, published, a few years ago, a case of a young girl who had applied *such* a pearl powder to her face for some time, and who consulted him for *great weakness in her wrists;* she was sallow, cachectic, emaciated, and very weak; suffered from loss of appetite, unpleasant taste in the mouth, constipation (but had no blue line in the gums), vomiting, lurid spots and superficial ulceration in the legs; the wrists dropped; power of extension of the fingers almost lost and the muscles of the fore-arm very emaciated.

Notwithstanding the internal use of *Iodides of potash* and *Iron, Strychnine,* and *Cod-liver oil,* and external friction, constant exercise, faradisation, and elastic supports for the wrists, the disease continued, and when the arms are extended the hands hang helplessly down by their own weight, and the muscles are shrunken; the sisters of this patient also used this face powder (carbonate of lead), but being in better health were less susceptible, and so escaped the effects of the poison. Children apprenticed to potters and house painters show also similar effects of lead poisoning.

44. *Mercurial paralysis*

is characterised by tremulousness and shaking, sallow, cachectic and emaciated appearance, ulcerated condition of the mouth and gums, looseness and loss of teeth. These symptoms are frequently observed in apprentices at gilders, mirror manufacturers, where the glass is covered with mercurial amalgam; it is also brought on by abuse of mercurial internal and external treatment, and by absorption and inhalation of mercury. Thirty years ago, when a medical student, I visited the mercury mines of Idria in Carniola, and there I had an opportunity of seeing the effects of mercurial poisons by absorption through the skin, and by inhaling the air impregnated with mercurial vapours. The mercury is sublimated in large vaulted chambers, and boys and girls used to sweep the walls to which the small globules of mercury adhered. The sweeping is done a few days after the sublimation, but the chambers remain still very warm, and this high temperature favorable to the continual mercurial evaporation is very injurious to those who are obliged to work therein. Silver and gold are extracted on a large scale by amalgamation with mercury in Saxony and Hungary, and there also mercurial paralysis is caused.

45. *Paralysis from bad habits, and too early and frequent abuse of the sexual functions.*

Youths and adolescents are more frequently than is

generally believed and admitted the victims of bad habits, and I seriously wish to exhort my younger and less experienced colleagues never to forget, while examining young paralysed subjects, cautiously, and without giving offence, to find out whether any of these abuses are not the predisposing cause of the paralytic affection. I have seen instances of little children of the tender age of seven being taught by their older playmates in school or at home to play with the private parts. That the continual practice is necessarily, in proportion to the frequency of its repetition, sooner or later followed by very serious consequences is evident. A certain shyness, trembling and shaking, general weakness, and pains in the head, eyes, and stomach, back, and limbs, loss of appetite, tendency to constipation, difficulty of passing water, coldness of hands and feet, depression of spirits, incapability of reading, fatigue after short walks, dragging of the legs, incapability of sitting up and turning in bed, are some of the symptoms preceding the actual paralysis. In some cases I have traced the bad habit to bad nurses, who, for the sake of keeping them quiet, teach even infants to play with parts that should not be touched. In sickly youths, with sedentary habits and over-stimulated imagination, sexual function is too early roused, and, according to circumstances, either self-abuse or frequent sexual intercourse is the result. While collecting notes for this paper I was consulted in four instances, one of insanity under the form of apathy, another of epileptic or similar fits, a third of mental depression and shyness of seeing strangers, and the fourth of incipient paralysis, in which I have traced the complaints to bad habits. Only in the first case I could not hitherto confirm my opinion, because the patient is scarcely able to understand my questions, and unwilling to answer them.

46. *Temporary paralysis*

by pressure on a nerve continued for some time.

It happens sometimes that children sleeping with the arms hanging over the back of a chair or of the bed, on awakening find both arms paralysed for some time; this loss of power is caused by prolonged pressure on the brachial nerves. Mr. Paget, in a lecture " On some Causes of Local Paralysis,"* mentions a similar result after tight ligature.

47. *Paralysis caused by cancer or tumours in the brain or spine.*

The symptoms are similar to those of cerebral paralysis, or of those of a pressure on the spinal cord, to which are added some symptoms dependent upon the seat of the cancer and tumour, and in all these cases the paralysis is *gradually* developed; a sudden invasion has been observed only exceptionally.

48. *Congenital paralysis.*

There is from birth an unhealthy cachectic appearance, deficient form of the head, idiotic or stupid expression, some cerebral affection, intellectual faculties deficient, sometimes strabismus; hemiplegia is the most frequent form, also partial paralysis under the forms of club-foot and club-hand. The most frequent deformity is pes equinus varus. The pes calcaneus is very rare, and the combination of this with the arched instep is never congenital.

49. *Hip disease.*

As some forms of partial paralysis of children may be taken for *hip disease*, I subjoin the characteristics of this complaint.

" The child carries all its weight on the healthy limb, turns the foot of the affected side inwards when walking, and the toes of that foot are on the instep of the healthy one while standing; striking the head of the femur against the acetabulum by a blow on the heel causes pain; there is also a fixed pain in the knee, heat in the hip-joint increased, and every active movement with that joint causes pain." (West.)

* *Medical Times and Gazette*, 1864, vol. i, 717.

50. *Indications for the Treatment of Paralysis, and of the Deformities caused thereby in Infancy, Childhood and Youth.*

The indications vary according to the constitutional state of the patient, the causes, duration, combination, period of the complaint, and of the subsequent deformities.

As a general rule, sound hygienic instructions concerning air, food, drink, clothing, exercise, physical and mental education, form the first and most important prophylactics of all complaints during the growth of the human body. It is to be hoped that more attention will be paid to the study of these subjects, which are a very, perhaps the most, important part of any curative treatment.

Cachectic patients must be constitutionally improved if their paralysed limbs are to gain strength; rickets, scrofula, strumous, and other constitutional diseases are, as far as possible, to be cured.

Although often unable to remove the cause of the paralysis, we must, before a plan for the treatment is devised, try to find out whether the origin of the complaint is in the brain, spinal cord, or a peripheric nerve, or in any other organ.

The previous chapter on diagnosis will be sufficient to enable a careful observer to find out the character of the complaint.

51. *Improper use of Orthopædic Instruments, and of Tenotomy.*

The atrophic and localising paralysis, with its subsequent various deformities, has been placed at the head of this section, because the knowledge of the treatment of this form of paralysis might enable young practitioners, as well as those who cannot make paralysis a special study, to prevent many a contraction and deformity, and to arrest, as far as possible, the progressing atrophy of the various tissues; the study of this treatment will prevent the majority of most conscientious general practitioners from placing their patients under the care of such orthopædic surgeons as are always ready with the knife in hand for a tenotomy, even when the slightest improvement cannot result from such an operation. It will also prevent them from sending blindly their patients to orthopædic and anatomical instrument makers, who, like any other manufacturers wishing to get rid of their instruments and supporting apparatus, have a machine in stock for every deformity. As there are medical men who have no knowledge of the mechanism required in each individual case, the instrument maker applies the same instrument (which each maker considers as the height of perfection) indiscriminately, and without the knowledge of the individual complaint, to every case which is similar in appearance. Thus the little patient, with either the tendon of the healthy muscle cut, or with an instrument which should but does not cure, or with both cut tendon and instrument, is left to grow while the paralysed muscles continue to atrophy from want of the necessary and suitable means. There are exceptions in both orthopædic surgeons and instrument makers. It is with much pleasure that I refer to the first introducer of tenotomy into England, who with praiseworthy sincerity, during the last few years, has preached against the abuse of the operation he once very enthusiastically advocated. Professional men should never make use of the instruments of such makers as give consultations on their own account, or who do not carry out the instructions given to them.

We do not send our prescriptions to such dispensing chemists as prescribe on their own account, or do not give the medicines we wish. The same rule holds good with regard to orthopædic instruments. The remote cause of such a state is the ignorance in which medical students have been left by their teachers, who should give them at least an elementary knowledge of the cases in which tenotomy is necessary, and of the mechanism of an orthopædic instrument when required in a given case. I will finish these remarks with the following:—" It is certain that in many cases very satisfactory results may be obtained *without tenotomy and without apparatus,* by means used with the view of bringing back power into the paralysed muscles. Electricity, movements of various kinds, shampooings and others; and my own experience has convinced me that the fact is not yet sufficiently recognised and acted upon."—Dr. Radcliffe.

"In all cases of paralytic rigidity or contraction *passive* movements of the limbs should be practised diligently. I have heard of very unexpected results thus being produced. Steady perseverance in this proceeding may enable us to dispense with division of tendons."—Dr. Handfield Jones.

52. *Indications for the Treatment of Localised Infantile Paralysis.*

1. Those who believe that an extravasation or exudation takes place in the invasion of the disease, wish that means should be used for its absorption, and for the diminution of the pressure on the spinal cord, whilst those who believe that a state of weakness and a lower degree of nutrition of the spinal cord cause the paralytic symptoms, advocate tonics of all kinds, medicinal as well as hygienic ones.

2. To raise the temperature of the paralysed parts.

3. That the power of the will should be made use of as soon as possible, in order to influence dynamically the nerves, and to rouse the action of the paralysed muscles by the most important functional stimulus of the will, which cannot be replaced by any other.

4. To arrest the constantly progressing atrophy as far as possible.

5. To prevent the contraction of the joints and retraction of the muscles, and, consequently, the various deformities caused by them.

6. To remove the contraction of the muscles, the retraction of the tendons, and the abnormal state of the joints where they have taken place.

7. When this result is obtained, to retain the natural position by all means at our disposal, and thus to prevent curvatures of the spine, deformities of the limbs, dislocation of joints, &c.

8. Every weak joint is to be individually treated, and all

movements acting upon such joints are to be employed in a suitable manner.

9. Every group of paralysed muscles, or single muscle is specially to be attended.

10. Every weak part of the body, as a single limb, the neck or trunk to be specially taken care of.

11. If some improvement be obtained, several previously weak parts of one limb are to be induced to combine for the execution of a special movement, during which they are assisted by another person.

12. That during the progress of improvement two or more limbs should combine for action, either with or without simultaneous action of the trunk.

13. That as the patient improves, the same movement of the limbs should be tried in various positions, to begin with the lying and sitting; afterwards, while the trunk, or the legs only, are supported in forwards, backwards, and sideways, lying positions; later, full-, long-, half-sitting, half-kneeling, and balance standing; standing in stride-, close-, step-, pass-, and other positions are used.

14. That movements which at first appeared to the patient impossible, later only possible with external assistance, should afterwards be done by the patient's exertion alone, and finally even carried out while the patient slightly resisted.

15. To improve the power of sitting, kneeling, standing, walking, running, jumping, and all other movements ordinarily required in the daily occupations of life.

16. That supporting apparatus, crutches, sticks, or retentive contrivances, should be gradually removed, and only the indispensable ones retained.

53. Treatment of the (atrophic and localising) Infantile Paralysis.

During the first or feverish stage the only chance there is, is to prevent the further progress of the disease, but unhappily the duration of the fever being very short, and the symptoms usually not observed, paralysis sets in suddenly, and only then is medical advice sought. Hitherto there has been no authentic case known where the regular and very serious course of the disease has been arrested in the first stage, as the cases of infantile paralysis reported as cured in the course of a few days have only been temporary or reflex paralysis, with abdominal or other functional derangement.

The old school recommends in the first stage antiphlogistic treatment, local depletion of blood on both sides of the spine, either by leeches, cupping, or scarification; mercurial frictions, cold compresses on the spine, tepid emollient baths; cooling purgatives, calomel in large doses, and, if any local irritation is believed to be the cause, lancing the gums, enemata, &c., are advised.

In all cases where an infant or child is feverish and very hot, it is desirable first to inquire what food it had taken, whether the functions of the intestines were regular, or if it had been exposed to a severe east or north wind, or had had a fall, or had been in any other way injured. To relieve fever and heat, every hour or two hours (according to the intensity of the symptoms) sponging the body all over, either with tepid or cold water, cold compresses on the top of the head and down over the forehead, so far that the eyes are covered, are useful; these compresses should be changed as soon as they are warm. During the first hour they should be changed frequently, afterwards not so often; if the bowels have not been relieved an enema of tepid water, or if this does not act, one with a little soap or salt mixed with the water is to be used. If the feet are cold, flannels dipped into hot water and well wrung out should be applied; the legs should be covered to above the knees; cold compresses and dry cupping on both sides of the spine, if there is any tenderness or pain; the cold compresses on the abdomen are changed only when very warm. Aconite in very small doses assists in the feverish state. Fresh cold water, alone, or with a little sugar and lemon juice, or barley or rice water sweetened (flavoured) in a similar manner, has been found the best drink in the feverish state.

As soon as the heat in the skin lessens, and traces of perspiration are apparent, all the cold water applications should be suspended, with the exception of the compresses on the head or spine, if these parts continue painful: no food should be given until the child repeatedly asks for it, and then it should only consist of milk, bread and milk, decoction of cocoa-nibs, stewed or baked apples, or prunes, and a small quantity of bread.

Convulsive fits produced by indigested food are removed by spontaneous vomiting, or this can be brought on satisfactorily by tickling the roof of the palate, by drinking warm water with some fat in it, or by a weak infusion of common chamomile flowers; only when these simple means fail recourse may be had to emetics. Exposure to cold also causes in children a feverish state, accompanied with perfectly white clayish stools, and tendency to vomit. Complete abstinence from food, and perfect rest in bed, soon removes this indisposition without calomel or purgatives; a drop of tincture of *Nux vomica* in water is also used every three hours.

Abstinence from animal food (with the exception of milk and one egg daily), regularity of meals, and proper attention to the skin during the period of dentition, moderate exercise in the open air and suitable clothing according to the seasons, are the most simple and most neglected means for preventing fits, fever, and other ailments in healthy children. After a convulsion even of the slightest degree, or after any feverish state, infants and children should be minutely examined in order to lose no time in case of a sudden invasion of paralysis.

54. *Treatment after the invasion of paralysis.*

Some authors believe that soon after paralysis has been observed, it is still possible to counteract the cause which, as they think, is extravasation, exudation in the spine, or congestion; for this reason, setons, moxa, active cautery, and the cauterisation with a red-hot iron wire applied to the paralysed part, are amongst the means recommended by the old school, to which must be added the external application of tincture of iodine, croton oil, flying blisters on the dorsal and lumbar part of the spine, inunctions of tartar emetic to produce irritation of the skin and to act as counter-stimulants and absorbents. Rosen's liniment, *Phosphor oil, Oleum animale Æthereum, Ammonia caustica, Spiritus formicarum, Tinctura cantharidis, Oleum æthereum Sinapis*, have been used externally along the spine, and similar means have also been applied to the paralysed parts to stimulate them to increased action. Internally *hydrojodate of potass, Infusion of arnica flowers, and cod-liver oil* have been prescribed. *Nux vomica* and *Strychnine* have been used to such an extent that the characteristic electric shocks of the whole body have been observed without causing convulsive movements of the body or the paralysed limb; notwithstanding the simultaneous hypodermic and external use of the *Extractum Nucis vomicæ spirituosum* with alcohol, *Liquor ammoniæ causticus* on the spine and lower extremities in the majority of cases, and especially in the more intense forms of paralysis, no change in the power of motion could be observed (Heine). Dr. West has also been disappointed in the effects of *Strychnine* in the (*atrophic and localising*) infantile paralysis. *Quinine, Iron*, and *Arsenicum*, are recommended by those authors, who consider these the best tonics.

Amongst the mineral waters resorted to are mentioned Bath and Buxton, Pfæffers, Wildbad in the Black Forest, Gastein in Austria, Abano in Italy, Pöstény in Hungary; the last two are well known by their natural deposits which are used either as mud-baths in which the paralysed limbs are immersed, or as cataplasms applied to the affected parts.

Also artificial mud-baths with iron as those of Marienbad and Franzensbad in Bohemia, baths with salt, gelatine, sulphur, malt, potass, formic acid, and pine-needles' decoction have been recommended.*

Sesquichloride of Iron, Phosphorus, Arsenic, Bichloride of Mercury, which last he believes to be antiseptic and tonic and analogous to arsenic, are prescribed by Dr. Radcliffe in con-

---

* *Antimony* in small doses continued for many months together, with purgatives and stimulating embrocations to the limbs, constitutes the best [?] treatment for [which?] infantile paralysis.—*Holmes Coote.*

gestion of the spine and myelitis, and as he believes that the atrophic localising infantile paralysis is similar to that of spinal congestion, I suppose the same medicines will be used by him. Brown-Séquard, who finds hyperæmia of the spinal cord in contradiction to Dr. Radcliffe's anæmia, recommends *Strychnia, Belladonna,* and *Secale cornutum,* and a position of the body on the abdomen or on the side with arms and feet hanging down, which is supposed to drain the blood from the spine.

"Various stimulants, both internal and external, have been resorted to and recommended by different authors. The Bath and Buxton waters internally and externally. *Rhus toxicodendron, Nux vomica, Arnica Montana, Raphanus rusticanus, Cantharides, Semen Sinapis, Opium, Valeriana, Camphor, Castor, Ether,* mineral acids, ammonia, lavender, blisters, the actual cautery, burning with moxa, galvanism, electricity, warm and cold bathing, friction, &c.; each of these remedies has been extolled by those who have treated upon their respective uses, and *each in its turn* has DISAPPOINTED THE expectations of those who have relied on its exclusive administration (Ward)."

55. *Medicinal Treatment of various forms of Paralysis by the Physiological School.*

The number of cases of so-called infantile paralysis reported is very scanty; I will therefore quote only, sub. I, the names of those medicines which in common with the old school have been recommended in the various stages and forms of paralysis; sub. II, those which are not or rarely recorded by the latter school.

I.	II.
Arnica.	Anacardium.
Argentum.	Angustura.
Arsenicum.	Baryta carbonica.
Bryonia.	Causticum.
Camphora.	Cocculus.
China.	Calendula.
Cantharis.	Carbo vegetabilis.

Colchicum.
Cicuta.
Conium maculatum.
Helleborus niger.
Ipecacuanha.
Kali hydriodicum.
Nux vomica.
Oleander.
Pœonia.
Phosphorus.
Rhus toxicodendron.
Secale cornutum.
Sulphur.
Sulphuris hepar.
Veratrum.

Cuprum.
Dulcamara.
Graphites.
Gelseminum.
Kali carbonicum.
Natrum muriaticum.
Nitricum acidum.
Platina.
Plumbum.
Sepia.
Silicea.
Staphysagria.
Strontianum.
Zincum.

56. *Domestic and Popular Means recommended for the Treatment of Paralysed and Atrophic Limbs.*

"The rudest tribes of savages are found to have their remedies and modes of cure, often rash, violent and injudicious, though sometimes discriminated with precision and adapted with dexterity and skill" (N. Chapman).

57. *Frictions.*

The ancients recommended in atrophy and paralysis of the limbs frictions with woollen cloth, as one of the most important means.

"In frictione spes plurimum consistit."—(*Paulus Ægineta.*)

Fomentations with warm sea-water.—(*Celsus.*)

Rubbing with cloth impregnated with the smoke of sugar, mastic (*Taxus baccata*) or moistened with warm brandy.

Rubbing with horse-radish and the common black radish (*Raphanus rusticanus*) shampooing or kneading the limbs, and rubbing them with the lather of soap as used in the Persian, Turkish and Eastern baths. (Sir R. K. Porter, *Travels in Georgia, Persia.*)

Flagellation, rubbing, and percussion with dry birch rods

with their leaves slit or steeped in a lather of soap. The same operations with a bundle of fresh nettles.

To cover the paralysed limb with cataplasms of warm solution of potash and of fresh heaps of ants.

To bathe in the sediment left after distillation of brandy, to expose the limb to the vapour of brandy yeast, produced by placing pieces of red-hot iron into a vessel filled with the brandy yeast.

To bathe in infusion of thymus serpillum, in decoction of the cones of pine trees, in decoction of bran mixed with mustard and salt, in warm bulls' or of calves' feet and heads of sheep, in an infusion of ants (formica); a bag is filled with the heap of earth containing the ants, hot water poured over it and the limb bathed in it.

To cover the part with hair or cat-skin, with oiled silk.

Boerhave observed a quack curing in many instances paralysed limbs, and finally found out that he used the pulp of horse-radish, which he applied to the part and covered with a hog's-bladder; a scarcely bearable pain was caused and relieved by butter; afterwards a cure was frequently performed.*

The sediment of wine is frequently used in France and Hungary as a stimulating embrocation; also stimulating and tonic baths, with salt, iron, sulphur, seaweeds, steam douches, are recommended as well as a mild hydrotherapeutic treatment.

The great effects of so-called animal baths are often mentioned, they are used by immersing the paralysed limbs into the still warm blood of the animal as soon as it is slaughtered, or the half-digested contents of the stomach of the herbivora are used; the natural warmth of the blood and the gastric juice as well as the half-digested gramina are believed to have great power in restoring the loss of power.

Amongst the applications of animal heat, I find the prolonged immersion, even for hours, of the limb in warm blood of an ox which has just been killed, or into the

---

* "Empiricus quidam curam omnium paralysium suscipiens applicabat semper unum remedium cujus effectus erat dolor intolerabilis; ægri ejulabant ac si locus carbone pestilentiali inureretur; dolore per butyrum sedato sæpe sanatio."—Boerhave, *Pralect. de morbis nervor.*, 1761.

stomach of a sheep or ox immediately after it had been slaughtered. A dove divided into two halves is applied to the affected part while it still retains its animal heat, also the still warm skins of hares, rabbits, cats, immediately after being killed, have been used for a similar purpose.

58. *Electricity*.

Electricity is considered by many as a specific. Duchenne declares that all infantile paralyses where the electro-contractility has been only diminished, have been cured completely and tolerably quickly without atrophy and deformity, when Faradisation has been applied a few months after the invasion of paralysis (*Electricité Localiseé,* p. 298). It is probable that these cases have been only reflex and temporary paralysis, which recover frequently without any treatment.

Duchenne's (to whose assiduity and genius electro-physiology owes so much) belief with regard to the power of electricity is so transcendent that he considers it to be able to create a muscle. These are his own words. " The fatty substitution is often very irregular, as proved by electromuscular and microscopic examinations. In such cases the *healthy muscular* fibre can become the nucleus of *new* muscular fibres, and even of *a new muscle* under the influence of localised Faradisation."*

Surely Faraday never dreamt that his induced current would be the generator of new muscle. This belief in new fabrication of muscle is based on the following case of partial paralysis, in which the majority of the muscles of an arm was atrophied ; for two years Faradisation was continued in combination with localised gymnastic movements according to Ling's system. The fibres of the internal part of the anterior third of the deltoid which contracted by electric excitation changed to as many fascicles, which by degrees increased in volume ; afterwards other fibres ap-

---

* The original, l. c. p. 303. Dans ces cas la fibre musculaire saine peut devenir le noyeau de nouveaux faisceaux musculaires et même d'un *nouveau muscle* sous l'influence de la faradisation localisée.

peared in the anterior half of the muscle and became centres of new muscular fascicles, but *none appeared in the posterior half.*—Duchenne, p. 305, l. c.

Duchenne makes use of Faraday and Ling as the progenitors of a new muscle, but gives credit *only* to Faraday. Is this right? Why does Ling not share in this triumph? This case proves that the anterior half of the deltoid was in appearance entirely, but in reality not to such a complete extent atrophied as not to be influenced by Faradisation and (as Dr. Radcliffe calls it) Lingism. How does Duchenne explain that the posterior half of the deltoideus was not *newly fabricated?* To what extent even such an accomplished observer as Duchenne can be misled, when his attention is directed to one subject only, will be seen by the following:

"Should even all hope of saving the muscles from complete destruction be lost, Faradisation will still be useful, and in certain cases be necessary to the development of the bony (osseous) system." He believes that the shortening of bones, which amounted in some cases to from six to eight centimetres, was reduced by Faradisation to two or three centimetres. He appears not to have had an opportunity for observing that shortening of bones does not always occur even in very severe cases of atrophic paralysis, without any application of Faradisation or Lingism. Amongst the cases which came under my notice, no cause could be either found for diminution of the length of a femur, tibia, or of all the bones of a foot, nor for the normal length of these bones, although the degree of atrophy and paralysis was not different.

Laborde quotes a case of a child with paralysis and atrophy of both legs; in the right the complaint was more developed with the beginning of a club foot on that side, to which Faradisation was applied for a whole year either by himself or by one of the externs, without any visible amelioration. I know of two cases who, for the sake of daily electric treatment, have been staying with a London surgeon and electric specialist for a whole year, both of whom are still more or less atrophied. One, a boy with the tibia considerably shortened; the other a girl eleven years old, from Manchester, about whom I was consulted in

March, 1869. She was paralysed when two years old, and has lost completely the use of one lower extremity; she was under electric treatment for several years, partly in England, and again in Germany, and under the care of eminent physicians, without any visible amelioration in the nutrition of the paralysed muscles. In the last-named case the parents have for the last four years engaged a female attendant, who daily for two hours manipulates the affected part, and applies, as far as she has been instructed, active and passive movements to the paralysed limb. The parents ascribe all the improvement hitherto gained to these active and passive movements.

Before finishing this paragraph on the value of Faradisation, I will add another case where Duchenne himself has applied it for *two* years, and where it has not even arrested the progress of the disease.

"The boy Philipot was sixteen months old, in perfect health, and began to walk, when, without any known cause, his limbs were paralysed. For eight or ten days before the invasion of paralysis he had a slight fever. The localisation took place in the lower extremities and afterwards only in the right leg; the foot turned inwards and the toes were directed downwards (equinus varus). Notwithstanding the two years of tonic and electric treatment under Duchenne, the progress of the disease was not arrested; three years have passed since the boy was paralysed; his right lower extremity is completely powerless; the paralysis affects especially the abductors and flexors of the foot; his club foot is an equinus varus developed to such a degree that the equinus forms a right angle; external force is insufficient to replace the foot in the normal state. The resistance of the posterior muscles of the leg is extreme. Reflex movements exist partly; no voluntary movements except flexion of the toes. The atrophy of the paralysed limb is considerable, and the size of the muscles is reduced to such an extent that the circumference of this leg is only one half of that of the healthy one. When touched the atrophied muscles are particularly flabby and flaccid. The bones are similarly reduced in size; it is easy to see that the right knee pan is one half the size of the left; the articular ends are likewise reduced. The muscles of the thigh are also atrophied. The right leg is one centimetre and a half shorter, which causes much limping whilst the child walks, which he does with trouble, and whilst throwing—if we may use the term—his leg by the action of the crural muscles.

"The temperature of this limb is very notably lowered, and the colour is slightly cyanotic, the child complains of cold, and the mother, who has observed this for a very long time, tries to lessen it by two or three pairs of woollen stockings."*

These cases prove sufficiently that there is no reason for placing such a high trust in Faradisation, and that it is neither a specific and scarcely a more powerful stimulant than the other means which have been externally used. It has the one and great advantage of permitting the local application on a single group, or even a single muscle. I am sorry to say that I have seen in numerous cases the application of electricity left to the tender mercies of an ignorant nurse, because the medical men prescribing electricity do not know much of the scientific application of localised electricity; it is at present still the general practice to inflict on the little patient as many electric tortures as he can bear.

At present a boy eight years old, whose legs are seen in Fig. 16, page 28, is under my care, who was Faradised during eight months three times a day for fifteen minutes; later, ten, five, and three minutes; there was not the slightest improvement in nutrition or power of the leg. In some cases more expensive galvanic apparatus (constantly advertised by the manufacturers as the great remedy in paralysis) are used in a rough manner, either by placing one pole in the hand, and the other in a foot pan filled with water in which the foot is placed. Sometimes one pole is placed on the nape of the neck and the other on the affected limb. Galvanic chains with the continued current cause frequently at their poles a burning sensation, and at present a case with the scars produced by the electric cauterisation of these chains is under my care.

* Observ. 50, Laborde, p. 217.

The really paralysed and atrophied muscles in children affected by the a. l. i. paralysis are very seldom influenced by electricity, neither by Faradisation nor by galvanic and continued currents, nor by statical electricity.

Although it cannot be denied that electricity in one or another form has been of some use in some forms of infantile paralysis, we have not yet precise data which are the so-called *proper* cases for application; notwithstanding my sincere wish to benefit my patients by any means, I cannot conscientiously say that I have seen any remarkable effects from electricity in the a. l. i. paralysis.

When electricity is used it should be applied locally on each single muscle or on the trunk of the nerve supplying a certain group of muscles; while the electric contraction takes place it is desirable to induce the patient to keep up the contraction by his will; my object in such a case is to bring the dynamic stimulus of the will to bear on the paralysed muscle. After the will had a slight influence on the paralysed muscles I have found in a few cases that Faradisation helped to recover the contractility of the muscle much quicker.

"*Electricity* is another remedy from which you may naturally anticipate much good; but the results which I have obtained from its use have, on the whole, disappointed me; often pain is caused without any or only temporary improvement. But though you must not be too sanguine in your expectation of good, electricity is one of the means that must not be left untried" (West, Heine). "As to the use of electricity which is now much in vogue I feel satisfied as the results of a large experience that it requires to be used with much caution. I have seen cases in which after the employment of electricity for some time that agent has apparently brought on pain in the head, and has elicited something like an inflammatory process in the brain" (Todd, in *Lectures on Paralysis and Diseases of the Brain*).

"In a few cases paralysed muscles which do not react to the most powerful induced currents, react energetically to a galvanic current of low tension slowly interrupted. This has been observed in facial palsy, in certain cases of infantile paralysis, and of local palsy, as in lead palsy of the extensors of the forearm and of other muscles, in paralysis of the deltoid, not from lead, in certain cases of muscular atrophy and in paralysis from traumatic injury of a nerve.

"Baierlacher and Bruckner in Germany, Hammond in New York, and Mr. N. Radcliffe, of London, have called attention to this fact" (Radcliffe). There is also an important objection to the application of Faradisation in infants, which is, that it causes great pain; it is not merely a strange sensation ("sensation étrange," Duchenne) which induces the little patients to shout, scream, cry and struggle, in order to escape the electric application. Only such children as are less sensitive, or can be induced by sweets and little presents to bear the pain, will be able to be regularly treated by electricity (Laborde).

Like any other stimulus, electricity can exhaust the vital powers of the part on which it acts too long or too violently; it is a known fact that the prolonged action of a *continuous current*, whatever may be the direction, is to render the nerve incapable of transmitting the action, and this power returns only when it is allowed to rest, or if acted upon by a current guided in a contrary direction to a current moving in the same direction; the action of the close continued current on the motor nerves differs according to the degree of intensity of the current.

Dr. Remak, of Berlin, has proved that all groups of muscles are brought into a state of tonic contraction by a continuous current acting on the motor nerve which animates them; he has also mentioned the wonderful anti-paralytic effects to be expected from the continuous current; in paralytic affections it is the practice to apply the induced current (Faradism) to a single muscle, and if the whole substance is to be brought into a state of contraction the electrodes are applied at the two ends of the muscle. It is believed that this artificial contraction and extension is a powerful agent for preventing the progress of atrophy; increase of heat is also proportional to the force of contraction and the length of time it continues; the difference of the increase of temperature in the skin covering the electrified muscles from the temperature of the surrounding parts amounts according to some authors to 5—6° Fahr., and the cooling takes place slowly and regularly, especially when the part is not exposed to the air. Dr. Remak's assertions regarding the effects of continuous currents were at first not confirmed by his experiments in France. But lately data have been made known that the continuous current applied at the two ends of the spine have caused not only arrest, but a cure of what has hitherto been called the *progressive* muscular atrophy : there are not yet any facts proving that in the atrophic forms of infantile paralysis the continuous current is of any use, excepting slightly raising the temperature during its application. Galvanic acupuncture and static electricity have been also frequently resorted to in paralytic affections, and when the body was impregnated with the electric fluid, sparks have been elicited from the spine and the paralysed parts.

No positive rules regarding the direction, intensity, quality of the current, the duration and repetition of its application have been given even by the most sanguine advocates of electricity and electric specialists; but they still agree that a single application should not exceed five to fifteen minutes, and that in wasting palsy, or in cases of advanced atrophy, a very large number of operations must be performed before the disease can be arrested.*

59. THE MOVEMENT CURE, OR MEDICAL GYMNASTICS.

Active and Passive Gymnastics.

There is not another more important means for restoring the power of paralysed muscles than to let them systematically and regularly be treated by exercise. In a word, gymnastics prove themselves to the fullest advantage, but it is necessary to restrict their use, which has not been sufficiently thought of by authors who advocate them in this complaint. (Laborde.)

---

* Further information on medical electricity is to be found in the works of Duchenne, Remak, Lehnhardt, Siemens, Tripier, Althaus, &c.

60. *Frictions.*

The manipulation of rubbing in various ways and directions on the surface of the body either with the hand or the fingers is called friction; the same term is also used when the manipulation is done with linen or woollen cloths, flannels, felt brushes and other materials and instruments.

Frictions act on the skin, muscles, joints, nerves, or vessels; this depends upon the manner in which they are applied; with the tips of the fingers, or with the internal surface of all the fingers simultaneously, with the palm of the hand or the whole hand: further, upon the amount of power and pressure used in these manipulations in which either one or both hands alternately are engaged.

Without reference to any substance used, friction itself is mentioned by all authors to be of the greatest importance; they say that the stagnant and slow circulation (which contributes very much to the progress of atrophy) is improved; that by preventing the stagnation of the blood, and by raising the temperature of the cold parts, it counteracts the incessant progress of the atrophy and changes the abnormal state of nutrition. Dry friction with the hand only or with hair and so-called electric brushes (these last made of vulcanised india rubber), with coarse flannel, exert not only a mechanical effect of stimulating the capillaries of the skin and minute ramifications of the cutaneous nerves, but they produce an electric process in the skin, and as the epidermis is idio-electric, it is a bad conductor of the electric fluid. In proportion to the amount of dryness of the skin and of the hand or material applied for rubbing, the quantity of electricity increases, the action of the nerves is more eliminated, the vigour, vitality and circulation of the skin is raised and the nutrition improved. Moist frictions do not produce much electricity, and this does not accumulate, but is absorbed.

Frictions with fatty and oily substances are called inunctions, and these were used by the ancients—at present olive oil, butter, and lard are mostly used either in small quantities with the intention of preventing the chafing of the skin, or in larger quantities with the intention of having them absorbed by the skin.

"Rubbing can bind and loosen; can make flesh and cause parts to waste. Hard rubbing binds; soft rubbing loosens; much rubbing causes parts to waste; moderate rubbing makes them grow" (Hippocrates). That a simple operation, which, notwithstanding the various methods of its execution, its combinations with other passive manipulations, and various medicinal substances, which has been indiscriminately designated by the name of rubbing, friction, shampooing, massage, or lately the *anatriptic* art, has always had its enthusiastic advocates amongst laymen as well as medical men, need not surprise. Like all enthusiasts they professed to cure complaints of all kinds, whether chronic or acute, paralysis, deformities, curvatures, tumours, sprains, fractures, dislocations, and a host of others; thus rubbing became a universal medium indiscriminately applied by the ignorant and uneducated, who frequently and justly boast of cures where eminent professional men have failed.

My younger colleagues will therefore do well to take the hint, and try to learn something about friction, its physiological and curative effect; thus, in a suitable case they will be able to give to the attendants and nurses the necessary instruction regarding the duration and the method of this operation; further, they will instruct them that the friction should be done in the direction of the afflux or reflux of the blood, or the efferent or afferent nerves, or on a special vessel, nerve, muscle, or joint, according to the effect they wish to produce; at any rate they will have the advantage not to be laughed at by a so-called professional rubber. As this is not the place for writing a treatise on friction, I give a list of a few works on this subject, which has always attracted the attention of medical men.

Adolphi, *De frictione.* Lips., 1707.
Dillen, *Frictionis usus medico-practicus.* Giess., 1714.
Wilkens, *Frictionum utilitas in medicina.* Lugd. Bat., 1716.
Vasse, *An frictus sit salutaris.* Paris, 1722.
Waldschmid, *De usu frictionum in medicina.* 1723.

Luther, *De usu frictionum in medicina.* Kilon., 1725.
Loelhæffel, *De frictione.* Ludg. Bat., 1732.
Hundertmark, *De singulari usu frictionis et unctionis.* Lips., 1740.
Assur, *De frictionis usu.* Hal., 1742.
Quelmatz, *De frictionibus abdominis.* Lips., 1749.
Louis, *Remarques sur les différentes espèces de frictions.* Ancien Journal de Méd., v. 207. 1749.
Kanis, *De frictionibus.* Vienn., 1756.
Jussien, *An otiosis frictio.* Paris, 1757.
Wesphal, *De frictione magno remedio anti-hypochondriaco.* Gryphisw, 1762.
Mellin, *Frictionum usus præstantissimus, &c.* Ten., 1766.
Brotonne, *An frictio sit salutaris.* Paris, 1782.
Delius, *De panni asperi lanei usu medico-chirurgico.* Erlang., 1786.
Baudry, *Sur l'utilité des frictions.* Strasbourg.
Winkler, *De frictionibus.* Ten., 1795.
Picnitz, *De frictionis et unctionis usu therapeutico et diætetico.* Viteb., 1806.
Cohen, *De frictionis usu apud veteres.* Berol., 1820.
Delamarre, *Sur les frictions sèches.* Paris, 1829.
Grossman, *De frictione medica.* Lips., 1834.
Balfour, *On friction.* 1819.
Cleobury, *Rubbing System*, by Mr. Grosvenor, of Oxford. 1825.
Beveridge, *The cure of disease by manipulation.* Edinburgh, 1859.
Johnson, *The Anatriptic Art.* London, 1866.
Laisné, *Du massage.* Paris, 1869.

61. *Gymnastic Treatment.*

"From the very first the efforts must be unceasing to bring the palsied limb once more into use, whilst, when the power is most impaired, attempts must be made, by the regular employment of passive exercise, and by friction of the limb, to prevent that wasting of a limb which is sure to follow on long-continued inaction.

"If a leg be affected the sense of insecurity deters a child from walking or even from making the slightest effort to do so, although carefully supported by its mother or nurse, and it will cry and refuse to make any movement. The attempts which evidently distress the child so are discontinued, and in the hope (too often a vain one) that in time the little sufferer will recover some power, much valuable time is lost, and the muscles waste, and permanent deformity of the limb results.

"In these cases two very simple means are often of great service in preventing the untoward occurrence. The baby-jumper exercises the legs most effectually, whilst, when there is even a very moderate return of power in the legs, the go-cart is of very great use, since it completely removes all sense of the risk of falling, and the little one soon begins to walk again. The go-cart has the disadvantage that it exaggerates the disposition to lean very much forwards in walking, which is observable in children for some time, even after they have learned to walk pretty well. The gait is thus rendered very unsteady. As soon as the child is able to walk in the go-cart, its use should entirely, or in a great measure, be discontinued, and substituted by a little jacket made of a stout material, lined and padded under the armpits, which is put on the child, a couple of straps of stout webbing, one end of which is fastened in front, the other on the back of the jacket; they must be sufficiently long to be conveniently held by an attendant, who supports (or rather partly lifts) the child, and prevents it falling forwards. The child, feeling perfectly safe, now perseveres in walking. Many of the worst consequences of paralysis are thus prevented, and a more speedy and complete recovery is obtained than could at first have been expected.

"If the child be five or six years old, and able to learn to walk with crutches, this should be done, because it will make greater and surer progress if entirely dependent on itself than if its weight is borne, or the possibility of falling prevented by a nurse or attendant.

"When the arm is affected the principles laid down are of the greatest importance. *Passive* exercise *must* be strictly carried out, the sound arm must be tied up, either altogether or for a considerable part of the day; coaxing, bribes, and all the inducements which move a little child's heart, must be brought into play as rewards for using the feeble limb. Raising a weight by means of a rope passed over a pulley is a mode of exercising the arms, which can be put into practice even in very young children, whilst in those who are older trundling a hoop with the feeble arm is a capital plan for joining work and play. I need not say that much care and patience are needed in carrying out any of these suggestions, and not a little of that intuitive love for children which teaches those who are its possessors how to extract fun and merriment from what might in other hands be a most irksome task (West). I may be permitted to observe that all those cases in which a go-cart, or the other means suggested by Dr. West, can be used, are not the really bad ones, and there is no doubt that in similar cases a considerable improvement or even a cure can be effected by perseverance during a long period.

"The *local* means for promoting recovery of paralysed muscles are possibly of greater importance than the general means.

"The efficacy of frictions and shampooings appears to be indisputable.

"The efficacy of proper movements can only be doubted by those who are unacquainted with the results arrived at by the movements cure and by systematic movements of one kind or another with or without the help of mechanical apparatus.

"Indeed it is to be hoped that the time is not far distant when a suitable gymnasium will be considered as much a part of the proper fittings of a hospital as the dispensary, and when medical men more generally will be alive to the importance of suitable gymnastics, not only as an educational but also as a curative measure. Surely there is a lesson to be learnt from the results of the carrying out of the movement cure, a lesson which the practitioners of orthodox medicine are not justified in continuing to decline to learn because it happens to have heterodox belongings.

"It is certain that in many cases very satisfactory results may be obtained without tenotomy and without apparatus, by means used with the view of bringing back power into the paralysed muscles. Electricity, movements of various kinds, shampooings, and others; and my own experience has convinced me that this fact is not yet sufficiently recognised and acted upon in practice. That in many cases neither tenotomy nor apparatus can be dispensed with, I fully believe; that in all cases the electrical and gymnastical parts of the treatment are of *primary* rather than of merely secondary importance I am every day more and more convinced, because each day I meet with instances of muscles which I should have once looked on as hopelessly paralysed being resuscitated by those means. Indeed I cannot but think that, so long as institutions especially set apart for orthopædic purposes are wanting in properly furnished electrical rooms and gymnasiums, there must be in some essential points a necessity of a greater reformation in orthopædic practice.

"There is reason to believe that the PROPER use of movements and manipulations will be of service in the treatment of *many forms of paralysis.*

"Orthodox medicine has much to learn from heterodox medicine in this matter, and it is to be hoped that no time will be lost in recognising this fact and in acting upon it. It is to be wished, indeed, that the results obtainable in paralysis and in many other cases by means of the treatment called Lingism (after its originator, Ling), or the Swedish system of gymnastics or kinesitherapy, or the movement cure, were more generally known, and appreciated at their proper value. Surely it is not right to refuse to recognise a truth because it happens to be presented in a manner which is more or less erroneous. Surely it is not right to neglect an important means of cure because many (certainly not all) of those who are alive to its merits and carry it out practically are quacks and impostors, and in alliance with other quacks and impostors. Surely the practitioners in orthodox medicine are *eclectics*, ever learning and ever bound to learn and apply every means of healing.

For my own part I may say that I have long been in the habit of using various movements and manipulations in the treatment of paralysis, and that I am every day more and more convinced that to omit such movements and manipulations in these cases is to deprive the patient of a most important aid to recovery."—Radcliffe, *Lectures on Epilepsy Paralysis,* 1864, p. 332.

"Of late years much has been said about the so-called Swedish exercises as a means of restoring the usefulness of paralysed limbs; and though, unfortunately, the direction of them has fallen into the hands of persons not the most likely to maintain the reputation of our profession, we must not on that account undervalue the benefit which they are capable of affording. Two principles seem involved in their employment: the one, the devising of such movements as shall best bring into play these muscles, the power over which is deficient; the other, the calling forth the active exercise of the will in determining them." (West.)

Besides constant warmth, localised galvanism, and guards against distortion from unbalanced actions of muscles, Mr. Paget recommends in local paralysis the diligent use, for years if necessary, of

Regular friction and shampooing, especially circular shampooing.

Constant voluntary efforts; constant endeavour to regain every lost movement; and when any such endeavour is effectual, frequent exercise of the recovered power.

Swedish gymnastics, *i. e.* set exercise for each muscle in which power is not wholly lost.—Paget, in a clinical lecture reported in the *Medical Times and Gazette,* March 26th, 1864.

I have quoted these extracts from the writings of Dr. West, Dr. Radcliffe, and Mr. Paget, to whom, as well as to many other eminent colleagues, I am indebted for kindly placing patients under my care, not only as a proof that the scientific application of curative movements is appreciated by unprejudiced observers, but also as an encouragement to those who, convinced of a scientific truth, and of its value to others, do not shrink from its public advocacy, although they expose themselves to be called quacks and impostors, by those who judge without examining, and in general by the ignorant. Thus Mr. Charles Dickens has given the sanction of his name for ridiculing the scientific use of curative movements, and for attacking my character, by permitting the insertion of a paper in the journal he conducts.

62. *The various means for raising the temperature of the paralysed parts.*

As long as the low temperature of the affected parts continues, the nutrition of the wasted parts does not improve, *but* the wasting will probably progress. It is, therefore, of the greatest importance to counteract the cold of the paralysed parts, which, as it has been mentioned before, predisposes to superficial ulcers of the skin, as well as to the arrest of development of all parts which are below the normal temperature.

The means applied for raising the temperature either convey, or aid in retaining a large amount of warmth, or they contribute to produce in the parts a higher temperature.

1. The *position* should be attended to in all cases; a paralysed part should not be permitted to hang down and to dangle about; it should be placed in a horizontal position, and the coldest part should be the highest, which assists the reflux of the venous blood.

2. *Clothing*—spun silk, a mixture of silk and wool; wool or fur garments should be worn next to the skin; it is only in exceptional cases that the hyperæsthesia of the cutaneous nerves does not permit any of these materials to be used. Here silk is placed next to the skin, and wool or fur over it. The paralysed part should be well warmed before it is covered with the bad conductors of heat.

3. *Dry heat* is conveyed by exposing the part to the direct rays of the sun; to radiating heat or a good chimney fire; for this purpose a screen is placed before the fire, and

through a hole in the screen the cold limb is exposed to the heat, which in this way does not inconvenience the rest of the body. It should thus remain until it is very warm; bags filled with, or baths of hot salt, flour, or sand, are also used, as well as vulcanised india-rubber bags filled with hot water.

The Turkish baths afford the means for conveying *dry* heat to a very high degree.

4. *Moist heat* is applied by means of fresh or salt water baths, by warm douches, showers, hot flannels and cloths dipped in the water, well wrung out, and applied to the paralysed part; oiled silk, or other impenetrable materials are placed over it.

The Russian bath is one of the best modes for conveying moist heat.

5. *Junod's exhausting apparatus.*—The necessary precautions with regard to warming the apparatus, and the limb which is placed in it, too great diminution of atmospheric pressure, too prolonged or too painful application, must not be neglected. A large amount of blood is retained in the capillaries of the skin, by which warmth and improved nutrition are in some measure promoted. The feeling of bursting or excessive fulness in the limb should always be avoided when this instrument is used.

6. *Electricity.*—The *continuous* current is applied either by galvanic chains, which convey a current with slight tension; also static electricity has been recommended for the same purpose. Dr. Burg's metallic bracelets and bands applied to the calf, thighs, and upper arms, are sometimes of use.

7. *Manipulations,* or various movements applied by another person on the cold parts of the patient; oscillation, vibration, kneading, fulling, rotation, percussion, pressure, palpation, tapping, chopping, longitudinal and transversal friction on a large or small surface, with the palmar surface of the finger, or the palm of the hand, or with the whole palmar surface of the fingers and hand; the so-called shampooing (massage) and its manifold operations belong to those means which not only produce an increase of

temperature in the cold limbs, but also directly contribute to the improvement of the nutrition.*

The Americans, who cannot spare manual labour, instead of professional rubbers, use machines. Dr. George H. Taylor, who has a movement cure establishment in 38th Street, New York, has lately sent me his pamphlet, in which drawings of an oscillating machine to be used by four persons at one time, of an apparatus for rubbing, vibrating, and kneading, are published. These machines, if they answer the purpose, might be recommended to the hospitals for the paralytic, where there are no means for the application of the manipulations required in every individual case.

8. *Active exercise* of all parts which can be moved, and especially of those which are nearest the paralysed parts.

9. *Applications of cold on the spine* in the form of wet or dry cold compresses, and of ice in the spinal bag, have also been useful in some cases, and were used by me for years before Dr. Chapman published his theory of the action of cold and heat on the spine in many cases of spinal irritation, with pain, and where cold hands and feet are some of the symptoms. I was guided by the pain in the spine, and sometimes have used cold and warm simultaneously on two or three different places, according as the patient felt relief by the application of the different temperatures; but I was not and am not able to explain why, for instance, a warm or cold compress in the middle, and two cold or warm ones above and below it, frequently produce considerable relief.

The following case of paralysis relieved by ice and friction is interesting, and was published by Surgeon-Major Broughton, of the Bombay Army, who was induced to make the trial in consequence of Dr. Chapman's publication.

"A poor girl, æt. 21, was paralysed and bedridden for upwards of a year; there was entire loss of sensation and

---

* Many of these manipulations have been described and engraved in my handbook on the *Movement Cure*, published by Messrs. Groombridge and Sons, Paternoster Row, where the reader will find the mode of their application.

motion up to above the knees; the feet were stiff and attenuated, but neither waste nor rigidity in the muscles of the legs; pinching and cutting the skin caused no reflex action; tenderness on pressure in the lower cervical and dorsal region; left leg an inch shorter, which was connected by an obscure (?) thickening of the sacral region. Health good, menstruation normal, no hysteria, no accident nor injury; has never played like other children, but was accustomed to limping upon one leg, and had a weakness in her back; this limp decreased as she grew older, and although not able to join in the sports of her companions, was able to go to service, and to perform her duties satisfactorily.

"The pain in the back, however, increased, much and varied treatment had been in vain resorted to, and she and her parents had given up all hope of recovery.

"I directed in October two pounds of ice in oiled silk bags to be applied to the spine every morning for two hours, followed by hand friction down the spine to the extremities for two hours, the whole body subsequently being cased in flannel.

"Great pain and distress was the effect produced; the skin of the part subjected to the ice was purple and congested, appetite impaired, bowels constipated. The patient was with difficulty persuaded to continue the treatment; at the end of ten days a little sensation commenced in the feet, soon followed by slight motion. On December 1st she was able to stand and to move about the room. On Christmas day she was walking in the streets of the village, and two months later was able to conduct the ordinary duties of a cottage household. *No medicines have been used;* the constipation was removed by wet compresses." (*Medical Times and Gazette*, 28th May, 1864.)

How much the frictions for two hours daily may have contributed to the cure Dr. B—— does not mention, and there is no reason to ascribe all the good effects only to the application of cold, especially as it is known that the advocates of animal magnetism have published cures of paralysis effected by mesmeric passes and frictions; and I have at present a little paralysed boy under treatment whose leg and foot, according to the mother's statement, began to be warm only after having been daily mesmerised for five minutes by passes on the paralysed leg and foot.

10. The apposition of a *mineral* magnet is also mentioned as producing warmth or increasing the temperature in a cold, paralysed part.

11. The friction of the paralysed part with a hand dipped in cold water till the water evaporates often causes a sensation of warmth and a kind of glow in the paralysed part; for the same purpose a towel dipped in cold water and well wrung out is thrown on the cold part, and the passive movement called "fulling" is continued till the towel and the cold part become warm.

According to circumstances the medical man must choose the means most suitable in each individual case for raising the temperature of the paralysed parts; my object in suggesting and enumerating the various means is to call the attention of the profession to the importance of raising, if possible, but at any rate of conveying and retaining, a higher degree of temperature in and near the paralysed cold part.

63. *On the influence of volition on the paralysed part, and how to stimulate the power of the will.*

Paralytic patients are apt to remain quiet and not exert themselves. The stimulus of the will already diminished by the disease is still further weakened by want of action, not only in the paralysed parts, but also in those parts which are still partially under the control of the will. Rigidity or relaxation of the joints, contraction of the limbs, atrophy of the muscles, anæsthesia or hyperæsthesia in one or another part, and other symptoms caused by the paralysis, increase in intensity as long as active or passive movements are not counteracting their further progress.

Besides the various manipulations mentioned amongst the means for raising the temperature of the paralysed parts, all the movements which can be made passively on any joint according to its anatomical structure will aid in arresting or in preventing the atrophy of the muscles, rigidity of the joints, and the consequent deformities; but the most important factor will always be active movement, because the agency of the will is the only direct means by which we can influence the central organs.

About ten years ago* I advocated the *use of the will*, the *innervation* which is an organic functional act subject to the laws of waste and repair of the tissue performing it, as a most important agent for regaining the power of movement in the paralysed part. My experience since that time has but confirmed my views on the subject, and I can only repeat that strychnine, electricity, and other stimuli, cannot be substituted for innervation, the organic influence of the will proceeding from the mysterious, internal, vital source; all paralysis capable of some improvement can be more lastingly benefited by natural innervation than by artificial, electric or any other stimuli.

"To restore a part to its pristine state of vigour we must attend to the principle, that *active* exercise of an organ is necessary not only *to its perfection but to its preservation.*" (John Shaw.)

"The manner in which stimulants are supposed to act in paralysis is by increasing that energy of the brain which is necessary to the production of muscular action. *The stimulus which appears to me the most safe, the most completely under our control, and the best calculated to effect this object, is that of frequent exercise excited by or dependent on volition.*

"The power of immediate association between volition and muscular action can be only recognised by repeated attempts; want of attention to these circumstances will explain the general failure of the usual means that have been resorted to for the cure of paralytic affections after the primary disease of the central organs, whatever may have

---

* *Contributions to Hygienic Treatment of Paralysis*, page 12. Groombridge, 1860.

been its nature, has been removed. The intimate connection and dependence which exists between the sensorial and muscular power has not been adverted to, and *that most powerful of all muscular stimulants*, VOLITION, has been altogether overlooked, or regarded only as a casual or secondary means of cure." (Ward.)

"Of the efficacy of the will, as a subsidiary means of restoring power to the partially paralysed limb, I have no doubt whatever. Of course in the child, whose will is feeble and liable to be distracted by very trivial causes, this power is far less energetic than in the grown person, but still it is a power well worth cultivating, and the steady perseverance in it exercised from childhood up to adult age will, I am sure, do more towards the recovery of a paralysed limb than would ever be imagined from its casual employment on one or two occasions." (West.)

To move and exercise the paralysed limb as much as possible, in order to direct *perseveringly* the *mental activity* on the paralysed part, and thereby to contribute to the restoration of the lost movement, is *John Hunter's* advice.*

Although all authors agree in recommending exercise for the paralysed part, there are but few who insist on the necessity of an enduring psychical tension on the part of the patient *to try* to make an effort of moving a part, notwithstanding they are conscious that they are unable to move. Although children under the age of four or five years are scarcely supposed to be able to understand how to make the mental effort of directing the will in a given direction, I have several times seen children only three years old trying to assist in moving a paralysed limb whilst it was, in fact, moved by another person. In older and more intelligent children, as well as in youths and adults, the difficulty of making an effort of the will decreases in proportion to their intellectual development and mental energy; a sufficient amount of difficulty and resistance to the mental exertion of energetically willing we must be prepared to overcome in *all* cases, in the young as well as in the old patient, but

---

* J. Clarke, *Comment on some of the most important Diseases of Children.*

if the medical man has sufficient energy, patience, perseverance, and knowledge how to manage the little ones and to make use of their power of imitation, and, I may add, if he has some love for them, he will finally succeed, and his hand will feel, before his eye observes it, that the child tries *to will*, and this is the first and a very important trace of real improvement. A loving mother and an intelligent painstaking nurse interested in the child's welfare are most desirable assistants. It is very well to recommend elastic bands with handles, all kinds of mechanical and gymnastic contrivances, such as go-carts, baby-jumpers, &c., for cases where partial power of movement exists, but these cannot be used when there is *an impossibility of making the slightest movement.*

The following is the manner I have hitherto found best answer the purpose of rousing the will and of increasing the innervation :—

1. The patient is placed in a comfortable position, lying, half lying, or sitting, in which he can remain without any exertion perfectly passive.

2. The paralysed limb is raised and supported by the operator, who moves it in a given position, bends, stretches, turns or rotates it several times, while the patient is requested to see the movement which is done for him.

3. First only one movement is done. Suppose the extensors acting on the elbow-joint are perfectly paralysed, and that the patient is able to make the flexion, the operator makes the extension, and after having done so several times, the patient is directed to stretch the arm. The usual answer "I cannot" is not permitted; he is induced to say "I will try," and is encouraged to try; but as the patient knows by experience that he is not able to stretch the arm, and it is probable that for some time he has made no attempt at trying to do it, he finds a difficulty even in making the mental effort of trying. After some persuasion and encouragement he gives way and endeavours to try. At this moment it is important to make for him slowly and gently the extension, which being coincident with his mental process is at the same time seen by the patient;

the impression thus conveyed through the eye to the brain convinces him that his effort was not in vain, because the arm is really stretched. The little patient believes that he has some share in it, because he knows he has made some effort, which is itself a new and unaccustomed sensation. These efforts of the will are repeated three or four times, and the corresponding movements are always done for him. I consider this mental dynamical operation the first step towards influencing the conducting or efferent nerves, and indirectly the paralysed muscles. When flexors and extensors are simultaneously paralysed, the operator will proceed in a similar way. In this case he is obliged to do alternately and slowly the flexion and extension of the forearm, while the patient endeavours to direct his attention and will to the movements as is suggested in the previous case.

When the joint to be moved is rigid, the attempts on the part of the patient's will should begin only after the operator has repeatedly and for some time endeavoured by passive movements, manipulations, and other means to *ease* the joint and undo the rigidity.

4. The greatest precision is desirable regarding the direction of the movement, its uniformity and duration.

5. The patient's will is to act simultaneously with the operator's will, or according to his command.

6. The repetition of the movement depends upon the patient's mental and physical powers.

7. The sympathetic influence of the will is made use of by inducing the patient to move also the *healthy* limb, while he tries to influence the corresponding paralysed part.

8. The parts adjacent to the perfectly paralysed part should be moved *actively*, i. e., by the patient alone, if this can be done, or while assisted by another person if the patient's power is not sufficient to overcome the resistance of the weight of the part to be moved; it happens frequently that the inactivity of a part continues only because the patient is unable to move it with ease, whilst a slight diminution of its weight enables him easily to overcome the difficulty.

9. The energy of the will is increased in parts under the

patient's perfect control, by another person gently resisting the movement he intends to do, or by the patient's gentle resistance to a movement of his body or limbs, while done by another person.

The following are the successive stages regarding the functions of volition on the part of the patient:

1. The patient is encouraged by another person to make the first effort to will.

2. The patient tries, his power of moving is deficient; the movement is done for him.

3. The patient exerts his will, has but partial power, another person assists his movement either by diminishing the weight of the part or by actually doing a part of the movement.

4. The patient's will and power are sufficient to move the weak part, although not yet with ease; his power increases by degrees and the weak part is moved with less difficulty.

5. The will of the patient is stimulated by another person gently resisting him, while doing an intended movement; or he gently resists the other person doing a movement with any part of his body.

Great caution is required not to fatigue the patient by the attempts to use his will; if a dull or acute pain in the head is caused by the patient's endeavours to will, these attempts must be stopped, and passive movements only should be used, till at a later period new trials can be made without causing headache.

64. *To prevent the contraction of the joints, retraction of the muscles, and consequently the various deformities caused by them.*

The contraction of the joints takes place when the flexors moving the joint in one direction remain healthy, whilst their antagonists are paralysed; secondly, when both flexors and extensors or abductors and adductors are paralysed, and the limb is left for a very long time in the same position, without any attempt being made to stretch the joint even *passively.*

In the first case the flexors have retained their natural power of contractility without being counteracted by their antagonists; they have a constant tendency to shorten, and in course of time to contract and retract to such an extent, that the bent joint cannot be stretched even by another person, except by using a great amount of force; here the contraction is almost a primary effect. Similar cases frequently occur in the cerebral forms of paralysis, and especially in hemiplegia.

In the second case, the perfectly paralysed limbs contract in a much slighter degree; according to the laws of gravitation the joint is constantly tending towards one and the same direction; the ligaments and one set of muscles in consequence partly of the elasticity of the fibrous tissue, partly by the constant repose, and again by the extremities of the muscles being mechanically approached by the constant and unchanged position, contract and become retracted. The following case will explain this second form of development of a secondary contraction:—A child affected with localised paralysis of the flexors of the foot is not attended for months or even years; the child remains either in bed or on a couch, the foot, with toes usually projecting beyond the support, always falls forwards, and mostly also inwards; it forms an obtuse angle with the leg; the ligaments in front of the ankle-joint are slightly stretched by the weight of the foot, the heel approaches the posterior part of the leg, the tendo-Achillis is mechanically compressed, retracts in course of time, and the gastrocnemii having retained their contractility, a club-foot *equinus varus* is formed.

65. *Orthopædic apparatus.*

The contractions, which soon after the invasion of paralysis are very slight, are prevented either by bandages and splints, or other mechanical means which retain the joint in a normal position *when the patient is at rest*. If the flexors of the foot only are paralysed, an elastic power supplying the lost function of bending the foot is adapted to the boot, which should be high enough to reach two or three inches above the ankle-joint. In very slight cases a piece of leather (a flap) fixed to the middle of the upper part of the boot, or two leather straps attached to both margins of the sole, at a place corresponding to the middle or anterior third of the foot, and crossing each other on the dorsum of the foot, are fastened by a buckle and strap (passing through a leather loop to retain the right position) on the back of the lower part of the tibia, or the highest and posterior part of the boot. When the extensors of the toes are paralysed, I make use of stockings provided with divisions for each toe (similar to the fingers of gloves), which are quite sufficient for retaining the normal position of the toes, and preventing this contraction.*

If the precaution of preserving the normal position of the foot has been neglected for a long time, and the tendo-Achillis is more contracted, a greater amount of power is required in order to retain the normal position; for this purpose an upright leg-iron, figs. 42, 43, b, b (with a joint h corresponding to the height of the ankle-joint), is fixed at a right angle to the sole of the boot, which is provided with a socket placed transversally under the anterior part of the heel; the transversal piece joined to the lower part of the perpendicular passes through the socket i, into which it is fixed by a spring (pressing on this spring permits the removal of the perpendicular from the boot); the perpendicular fixed at the top by a padded ring d, which surrounds the upper part of the calf, serves only as a point of support for either a cord of vulcanised india rubber, or a spiral steel spring c, c, which is fixed at its upper part, and acts with its lower free part on a little wheel a, on the ankle-joint, and raises the front part of the foot into the natural position.

The patient having retained the power of the gastrocnemii, is *not* prevented by this elastic contrivance from stretching the foot, and when the patient is passive, the spring retains the normal position of the foot, acts as an antagonistic power to the gastrocnemii, which, although able to contract at the will of the patient, remain in a

* I fancied I was the first to invent similar stockings, but in 1862 I saw several officers belonging to the Japanese embassy who had stockings with a division for the big toe.

relaxed position; thus the ankle-joint retains its normal state, and the development of a pes equinus is prevented.

The Figs. 42 and 43 show the boot with the spring fixed

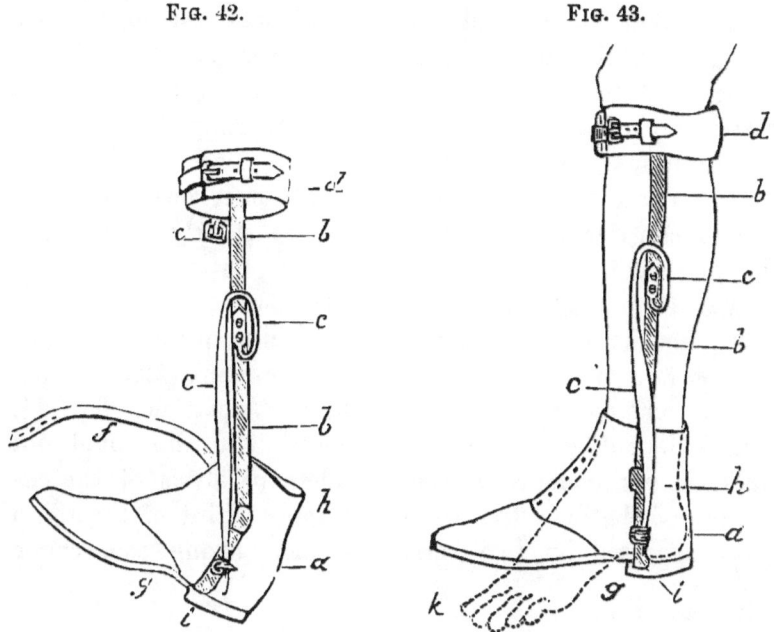

FIG. 42. FIG. 43.

to the perpendicular, and the position of the boot which forms an acute angle with the perpendicular in fig. 42.

Fig. 43 shows in the dotted lines the pes equinus, and how the foot is retained in the normal position.

If there be a tendency to varus, the spring will be everted and combine the functions of flexors and abductors, or in a slight degree a flap or strap is adapted to counteract the constant tendency of the healthy muscles to move the foot in the direction of their action.

66. *How mechanical supports should be made and used.*

The principles according to which every mechanical contrivance required for the retention of a joint in its natural position should be made are—

I. Not to interfere with those movements which the patient can do.

II. Not to interfere with the circulation of the affected or any other part.

III. Not to cause any swelling or pain.

IV. The elastic power, whether india rubber, steel spring, or any other material, should be sufficiently strong to act antagonistically to the prevalent power of the healthy muscles, which must be neutralised while the joint is at rest; they must be prevented from contracting except under the additional impulse of the will.

V. Where the tendency to contraction is very great, the elastic contrivance should be used constantly during the day and night.

VI. It is frequently necessary to remove the apparatus, partly for the application of baths, douches, passive manipulations, &c., or if they cause soreness, pressure, &c.

Mathieu constructed, in 1852, an apparatus, fig. 44, with india-rubber bands, for a young man who could not walk, in consequence of the complete paralysis of the extensors of the leg; the leg to the upper third of the thigh was supported by two vertical irons; strong caoutchouc

FIG. 44.   FIG. 45.

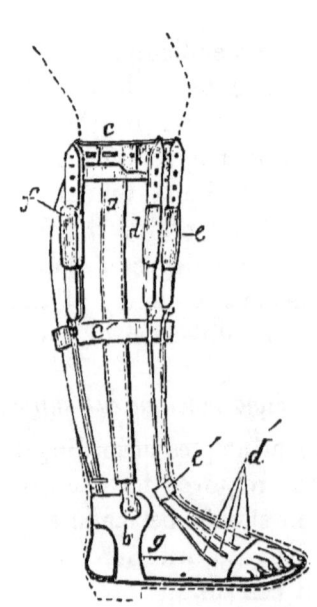

bands *a, b*, were placed in front of the knee, and fixed at the upper and lower circle; under the influence of the flexor muscles, the knee was bent, and the elastic band slightly stretched. When the action of flexion ceased, the band extended the knee; thus the movement of flexion and extension was alternate, and the patient able to walk; walking was at first difficult, but practice made it afterwards easy.

Fig. 45.—This is an apparatus by Charrière, according to Duchenne's directions, and consists of two uprights, *a* (which are united by the two metallic circles, *c, c*, which can be opened in front), fixed to the metallic piece, *b*, to which is attached a leather, or thin metallic plate covered with leather; *d, e, f*, are caoutchouc band substitutes for the paralysed muscles, and are fixed like tendons to the gaiter, *e', d'*, over the foot; the dotted lines show the outlines of the boot.

67. *Means for arresting the constantly progressing atrophy.*

The majority of means suitable for increasing the temperature of the paralysed limb are also used for preventing the progress of atrophy; besides these, kneading, pressure, and percussion, the continuous electric current, local inunctions, that is, frictions with oily or fatty substances (lard or olive oil are usually chosen), for one to two hours daily; both hands are alternately used, and the direction of the friction is longitudinal towards the periphery, or circular; the manipulations act on the capillary vessels, which supply partly the functions of nutrition, impaired by the abnormal state of the central organs; thus the pale and flabby parts improve in colour and firmness, and the atrophy is usually arrested as long as the paralysed part is daily acted upon.

68. *To remove the contraction of muscles, retraction of tendons, and the abnormal state of the joints.*

Neglect in preventing the secondary effects of paralytic affections of the limbs by all the means which have been named under the preceding head obliges us to remove them by means which are more difficult than those required for their prevention.

The preliminary treatment is longer, and consists in using all those manipulations of passive flexion and extension, or adduction, abduction, and rotation by which the contracted or deformed joint might still be placed in a normal position. If these movements leave no hope for an improvement, recourse is needful to extension apparatus adapted to each joint, and the power of which can be by degrees increased. In the majority of paralytic deformities these extension apparatus and the manipulations will be sufficient. To return to the previous instance of a *pes equinus:* when this reposition of the foot into its normal position is very difficult or can only be accomplished by the use of much force, *Scarpa's* boot, or any of the modifications of Delpech, Venel, &c., are used. After the lapse of many months, or even a year, the desired result is obtained, and then only can the retentive elastic contrivance be used.

69. *Tenotomy.*

Only in the highest degree of retraction of tendons and of deformed joints where no other means leave any hope for restoring the normal position of the joint, and if there be sufficient reason for expecting any practical result and benefit for the patient, tenotomy might be used; but we must bear in mind that the cutting of the tendons of the healthy and contracted muscles will not restore the power of the atrophied muscles. The abuse of tenotomy in paralytic clubfoot and contractions of the knee-joints has been, and is still, very general. To cut the healthy although contracted muscles, or rather their retracted tendons in a half paralysed limb, is very *irrational.* Ten years ago I advocated similar views, which have been confirmed by Bouvier and Malgaigne,* who say " the abuse of tenotomy in the treatment of clubfoot has been pushed to the highest possible degree, to the great

* " L'abus des sections tendineuses dans le traitement du pied bot, a été poussé aussi loin que possible au grand et irreparable prejudice des fonctions du pied que l'on avait la pretention de ramener à son état normal."—Malgaigne, *Leçons d'Orthopædie,* p. 118.

and irreparable loss of the functions of the foot; by which the advocates of this operation *pretend* to restore it to its normal state."

As a rule tenotomy is useless in paralytic deformities, except for changing the form of the joint without an improvement in the use of the limbs. In extreme cases of contraction which have not yielded to the protracted and constant use of passive manipulations and extension apparatus, and where the restoration of the normal position of the joint can be of any use to the patient, the operation might be performed, but also in such a case the smallest number of tendons or only one should be cut. It is unhappily the prevailing fashion of orthopædic surgeons to operate as often as possible, and cut as many tendons as appear retracted.*

Even in considerable contractions of the healthy muscles, the patient retains the power of still more contracting these muscles. He should be encouraged to do this while the operator simultaneously tries to stretch these parts by causing an excentric contraction of the muscles, which means that the two extremities of a muscle, while contracted by the will of the patient, are removed from each other by the will of the operator. This process frequently repeated produces results which are scarcely expected in the beginning; whilst this excentric action takes place, vibration, kneading, percussion is made on the contracted tendons.

* This mania for useless operations is proved by the following case of one of my patients, who, after having been cured from some ailment, wanted to get rid of an anchylosis of the second joint of the big toe, the consequence of an accident and bad treatment. Once I happened to mention that excision of joints is sometimes resorted to, but that this operation would not be suitable in his case, and that I could not conscientiously advise it. A few months later an orthopædic surgeon had performed this excision of a wedge-shaped piece of the anchylosed part, and the patient had been afterwards several times under the influence of chloroform, in order to enable this orthopædic surgeon to make passive manipulation on the operated big toe. The result was that the operator received his fee and the patient has still an anchylosed toe. Similar cases of cut tendons, without any beneficial result, I have repeatedly seen, and at this moment I have three paralytic patients who have undergone such operations without any beneficial result.

70. *To retain the normal position of the joints and to prevent curvatures of the spine and deformities of the limbs.*

The first and second part of this indication is fulfilled by the means named under the heading of 64, page 92. I will only add that the *elastic force* was first used in orthopædic apparatus by Delacroix, but Mellet was the first who used india rubber; and Rigal de Gaillac to substitute the elastic force to the weak muscles. Duchenne is believed to be the first to substitute or replace the paralysed muscles by artificial elastic bands or elastic springs (as they have been and are still used in braces) which are attached and placed in the direction of the affected muscles, and are provided with artificial tendons sliding like the natural ones in their proper artificial sheaths.*

Mathieu, the well-known manufacturer of surgical instruments in Paris, was the first who made in 1852 an apparatus of strong bands of india rubber, which are placed in front of the thigh and knee to supply the lost power of the extensors (fig. 44). Mathieu asserts that Duchenne has no share in the invention of the application of india-rubber bands except in having given him in 1853 the advice of using it in spinal deformities ("deviations de la taille").

The third part of the sixth indication, viz., *the prevention of deformities of the spine,* is carried out by attention to the position of the patient, who should always either recline or at least be supported while sitting. If one arm is paralysed it must be raised and the patient not be permitted to use

---

* A few years ago a London surgeon, who carried out Duchenne's ideas by using elastic india-rubber cords, which he inserts and attaches at the corresponding places of the paralysed muscles by hooks soldered on flat pieces of tin, and which are fixed to the limbs by circular bands of sticking-plaster, communicated his mode of proceeding or his invention to one of the London medical societies. The report was scarcely published in one of the leading medical papers when an orthopædic instrument maker sent to the same paper a long reclamation against the surgeon, and promulgated himself as the inventor of the application of the vulcanised india rubber for orthopædic purposes. The partiality of the editor for the instrument-maker was proved by a refusal of the insertion of the facts I have recorded in the interest of those who really have been the first to apply elastic force in orthopædic treatment, of which important progress there is not yet sufficient appreciation.

the spine as a medium for throwing about the weak arm ; when one leg is paralysed and a supporting apparatus is used, *two* sticks are more serviceable and better than one when he attempts to walk.*

The ninth, tenth, eleventh, twelfth, thirteenth, fourteenth, and fifteenth indications will be best carried out by the movement cure, which is " surgery without a knife, and which is, in the hand of God, the human hand permitted now through insight into God's law, to be the saving instrument of earthly life and organization" (Bauman in his *Address on Surgery*).

Those who have taken or will take the trouble of following Dr. Radcliffe's advice, mentioned in the previous pages, regarding the study of the movement cure and its results in paralysis, will soon be able to appreciate its value and to make use of it.

I should be obliged to copy my previous treatises on the use of *Movements in Chronic Diseases*, and the *Handbook of the Movement Cure*, were I to give detailed descriptions of all the manipulations and movements required for carrying out these indications.

A knowledge of the importance of medical gymnastics for diagnostic purposes in many purely so-called internal as well as surgical complaints will be also appreciated, and the student will find himself possessed of a very useful curative accessory, which does not exclude any other hygienic, medicinal, or surgical means. It might be worth his consideration that he will be able to cure many a chronic complaint or deformity which he would fail to cure without this system.

71. *That supporting apparatus, crutches, sticks, or retention contrivances should be gradually removed, and only those which are indispensable retained.*

The progress of the patient's improvement will enable us to carry out the above-named suggestion, but care must

* More on this subject will be found in my *Prevention of Spinal Deformities*, &c., published by Groombridge and Sons.

be taken to seize the right moment. All patients accustomed to some mechanical contrivance are with difficulty induced to give them up; they are not sufficiently aware of their own power, and have neither courage nor self-reliance. Notwithstanding this, it is our duty to insist upon their giving up mechanical aids as soon as we are convinced that they can do without them.

72. *The forms of paralysis in which the movement cure, including all other suitable means, can cause an improvement or cure.*

1. Where no change of structure has taken place in the primitive fibres of nerves.
2. Where the (gelatinous transparent) marrow in the nerve tubules has not undergone any or only a very slight change.
3. Where the single nerve-cords have not changed their structure, or only very partially or slightly.
4. Paralytic affections which are caused by venous congestion, or stasis in the brain, spinal cord, or single nerve-trunks.
5. Paralysis after slight degree of extravasation of blood, or serous exudations into the brain or spinal cord.
6. Where the muscular tissue of the affected part has not been considerably changed by deposits of plastic lymph, or granular or fatty degeneration.

In all these cases, an improvement, or a cure of the general symptoms, and especially of the power of movement, may be reasonably hoped for.

Sometimes spinal paralytic affections are more easily cured by medical gymnastics than cerebral ones. Affections with predominant spasmodic contractions of single groups of muscles are also sometimes more easily treated gymnastically than those with predominant relaxations.

Hemiplegia and paraplegia are better managed by the gymnastic treatment than general complete paralysis.

The prognosis is better in cases when the muscular activity increases the secretion of cystoblastem, by which the morbid muscular tissue is changed, reconstructed, and a new formation of muscular tissue takes place, and where the innervation of the paralysed muscles can be improved. In the contrary cases the prognosis is worse. What has been mentioned is not to be considered as an apodictic fact, but experience must confirm, explain, and enlarge these views.

Medical gymnastics will be as useless as any medicinal or other treatment in paralysis with change of structure in the nerves, with considerable sanguineous or serous exudations, softening, suppuration, tubercles, tumours, cancer in the central nervous organs, sclerosis of the spinal marrow, thickening and fungous derangements of the cerebral and spinal membranes, exostosis, and other incurable changes. (Melicher, *First Report of his Medico Gymnastic and Orthopædic Institution.* Vienna, 1853.)

73. *Effect of the Movement cure on the paralysed.*

It may be mentioned in general that a paralysed patient, having been previously treated medicinally without deriving any benefit, when left to himself and remaining passive usually gets worse.

The scientific application of the movement cure will in a curable case either arrest further progress of the complaint, improve, or cure the patient. In the majority of cases the constitutional state, the usually prevalent venosity, the respiration, circulation, and digestion, will be improved, and the tendency to fatty deposit or the atrophy of the limbs will be diminished, the patient will mostly improve in some respect, rarely get worse, and *never* be injured by the treatment, if judiciously used.

The object of the treatment by movements is to rouse and stimulate the diminished vitality, to increase the organic function of innervation of the paralysed part of the brain, spinal marrow, the diminished or deficient innervation of the nerves or single parts of nerves, to restore the mobility and equilibrium between the activity of the paralysed groups of muscles, or of single muscles, and the greater power of their healthy antagonists.

74. *Conditions desirable for paralysed patients while under the treatment by movements.*

Besides the strictest attention to all hygienic prescriptions a certain amount of perseverance and endurance in the treatment is required, the patient must exert himself as much as possible to assist the treatment by directing his mental influence very frequently to the affected parts; by making use of the powers left in and near the paralysed part; by trying to pay more attention to these parts so as to increase and develop his powers of will, which being fallow and latent are roused, stimulated, and increased by the movements, which thus contribute to greater vitality, activity, and more power of moving the locally affected parts, as well as the rest of the body. The success of the treatment will be in proportion to the energy of the patient's will, his mental powers, bodily activity, and perseverance in carrying out the given instructions.

Experience teaches that the lethargic patient without will, power, and movement, by degrees awakens, shows traces of will and action, begins to be conscious that he is able to try to do something, observes that his endeavours are not fruitless, that some (although very slight) movement in the paralysed part is visible, that his hope for recovery is greater, his will is more energetic, his endeavours are more frequent, his power is greater, his perseverance continues, whilst his general health and constitutional state improve, his bodily functions become more regular, and he is happy to find that his limbs gain more power of action and increased movement. The medical man, whose patience, energy, and perseverance have also been tried for months and even years, is finally rewarded by the pleasure of having obtained a result which is entirely due to his own exertions, and which cannot be obtained by the frequently invoked *vis medicatrix naturæ*; he will thus prove the truth of an old saying, "THAT FOR MAN IS NO BETTER MEDICINE THAN MAN."*

---

* Peter Blesensis, *Liber de Amicitia.*

I. *Cerebral paraplegia, with lumbar kyphosis, contracted knees and talipedes equini.*

Miss ———, four years old, was placed under treatment about two years after the invasion of the disease. Her head was large, and fell forwards between the shoulders; the neck was scarcely visible. The spine was considerably curved, particularly in the lumbar region, where the kyphosis was very prominent; the hips contracted, and the knees bent; both feet affected with talipes equinus; the legs were emaciated, pale, and cold. She was unable to extend the knees, to bend the feet, and could not stand. After three months' treatment she could stand, and walk a little with the aid of the nurse, and the improvement a month later was such as to lead to the opinion that she would soon move more freely. Dr. West, who had previously seen her, expressed himself highly satisfied with the improvement, and urged the mother to continue the treatment, which consisted in the application of elastic supports to the feet, various manipulations on the thighs, legs, and feet, of movements of these parts, and of the spine, and also the use of salt water baths. Her general state of health, the power of holding up the head, of sitting and standing upright, and also her general appearance are very much improved; the legs have gained in size and firmness, the skin has a better colour, and the temperature of both limbs has considerably increased. She is now able to make use of the go-cart, which was previously quite impossible. The treatment was interrupted by an attack of whooping-cough.

II. *Paraplegia, with talipes equinus.*

Master B—, eight years old, was sent to me by Dr. Dudgeon, who attended the child during an attack of febricula, which for three or four days produced no serious symptoms. When recovering from the apparently very slight indisposition, he lost the power of raising the right thigh, extending the right knee, and bending the foot.

Although there was also loss of power in bending and extending the left foot, he soon recovered the use of the left limb, but the right leg continued powerless notwithstanding the application of electricity. He was under treatment for seven months, and at first he was not permitted to walk any distance; the limp was considerable, because he had to throw the leg from the hip forwards. His spine was very weak, and slightly laterally curved, the right thigh less voluminous than the left. Besides the appliances of the movement cure, he was sent to the swimming baths in tepid sea water, and douches were applied to the back and legs.

He can now raise the right knee freely, and the power in that foot has returned to such an extent that he is able to balance himself on this leg while the arms or trunk are moved in various directions. When his attention is fully directed to his leg he can walk sufficiently well not to show any limp. In this case the elastic support for the foot was used only after he had considerably improved, when he was also sent, at my suggestion, twice a week to the riding school.

III. *Paralysis of one leg, with talipes equinus varus, and weakness of the other foot.*

Master ———, eight years old, was sent to me by Dr. McKern. When about four years old was a well-developed boy, but through an attack of gastric fever the right leg and foot became paralysed. Three years later, when placed under treatment, the right leg from below the knee, and also the foot, were in a complete state of atrophy. Fig. 16, page 28, shows the legs of this boy while sitting. He was unable to bend the foot, which was affected with talipes equinus varus. On account of the weakness of the leg he could not stand on it; there was a slight lateral curvature; the head was much bent forwards; his shoulders raised and rounded; the chest flat. When walking the limp is very considerable. The atrophy is shown in the diagram—leg and foot are cold and flabby. After nine months' treatment he was able to raise and bend the leg and foot; the spine is straight; the chest well developed; he can stand on the paralysed leg, and while on it do several movements with the arms and trunk. He can walk with scarcely any limp when his right arm is raised. Although his leg has not increased in size, it has become much warmer and firmer, and he is now able to run. Should the treatment be continued there is every reason for hoping that he will be able to use the leg and foot quite freely, although the size of the former is at present much smaller than that of the left leg, and the development of the latter is slightly retarded.

Inunctions, the movement cure, cold douches of salt water, preceded by warm ones, bathing in tepid sea water, and an elastic support, have been the principal means used. This is the case to which I referred (page 28) in the chapter on electricity, which had been used for eight months daily without any benefit to the patient.

IV. *Paralysis of the left arm, and paraplegia, with left talipes valgo-equinus.*

Miss ———, fourteen years old, was directed to me by Dr. Gully, of Malvern. When eleven years old, was being tossed to and fro like a ball by two adult friends of the family, when one of them missed catching her, and she fell on her back from a height of from six to eight feet, in consequence of which she became generally paralysed about forty-eight hours after the accident. As soon as the acute symptoms consequent on the external injury had disappeared, Dr. Brown-Séquard and other eminent medical men were consulted, who advised tonics, electricity, &c., to be used. After the lapse of three years, as there was no visible improvement, she was placed under my treatment for about two or three months at various intervals. The paralysis was localised in the two lower limbs and the left arm. These three extremities were considerably smaller than the well-developed right arm and trunk. The power of extending the legs and feet as well as the left arm was very restricted. The temperature of the affected parts considerably diminished. There was considerable difficulty in walking, which was a kind of waddle, in which the alternate lateral movements of the trunk—similar to a pendulum—prevailed. She was generally weak, and had a lateral curvature of the spine, with the usual concomitant symptoms of round and high shoulders, flat chest, &c. When first seen she was unable to rise from a kneeling or sitting posture, except with the help of the right arm. She is now about eighteen years old, and, according to her own statement, able to walk from two to three miles without assistance; her general health is much improved; her left arm is much stronger, and she is able to use it freely.

Since the above was written I was agreeably surprised to hear that at the ball given in honour of her first appearance, she was able to dance quadrilles and lancers, and that it was scarcely observed that her movements are defective. Although not yet able to jump with the left leg alone, she can walk very steadily for a short time, and I do not doubt that she will still continue to improve if she carries out with perseverance the given instructions and prescriptions.

V. *Cerebral hemiplegia, with deformity of the spine, of the left arm, hands and fingers, complicated with epilepsy.*

Master ———, nine years old, was paralysed at the age of five. His head is large; face pale, eyes prominent, and surrounded by dark circles; the left arm cannot be raised, it is contracted in the elbow and wrist-joint, while the hand and fingers are deformed and atrophied, and with the forearm twisted inwards, the spine is considerably curved in a lateral and posterior direction (a combination of scoliosis and kyphosis). The left leg is very weak, the knee cannot be extended, the foot cannot be bent; along one side of the chest and the abdomen the cutaneous veins are varicose, and project in serpentine lines. Mr. Prescott Hewett advised the father to place the boy under my care.

After he had been about eighteen months under treatment the power in the leg had improved, the contraction of the arm diminished, and his general appearance much changed. At that time he was without any known cause seized with an involuntary movement, similar to blowing of one side of his lips and one check. These attacks increased in intensity, and first " le petit mal," and finally regular epileptic seizures occurred once or twice a week, during which the left arm and leg were considerably more convulsed than the corresponding parts on the right side. Notwithstanding the use of various remedies, including the bromide of potassium, which was given in large quantities, and continued for several months, the epileptic fits continued as intense and frequent as before, till finally one fit occurred which was the most intense he ever had; since that time no fit occurred. As long as these fits lasted scarcely any further improvement took place; but as his treatment was from time to time continued, the previous improvement was not entirely lost.

In order to recruit his general strength, I advised his being sent in the country, where he remained for a year. Unhappily, the manipulations and movements which I had prescribed were neglected, and when he returned to town he was, with regard to his deformities and the use of his arm, worse. During the last six months he was placed under a regular course of treatment by movements; his left arm was daily rubbed for half an hour, various contrivances for retaining the normal position of the contracted wrist-joint were used, besides a glove with artificial tendons, as suggested by Duchenne in his *Orthopedie Physiologique* (page 24). I had to invent some contrivances, which Mr. Ernst executed according to my suggestions, to my perfect satisfaction. The general hygienic treatment was not neglected.

The following is the result obtained up to the present time. The boy, who is now thirteen years old, can walk, jump, and balance on the left leg; his shoulders, previously always raised, are in their natural position; the spinal curvature is very much improved; he has the use of his left shoulder and arm; can climb forwards and backwards on a vertical ladder.

Although the contraction of the wrist has almost disappeared, and notwitstanding the power of bending and extending the fingers, he cannot keep the hand extended, nor abduct the thumb. The form of his drooping hand is seen in Fig. 39, page 41. His general appearance and form has considerably changed. When the hand is supported by the mechanism adapted for that purpose, he is able to make partial use of his hand, the fingers have gained in size, and have a more normal appearance. As the manipulations and some active movements are still carried on daily, his improvement is still progressing.

VI. *Paralysis incipiens.*

Master ———, fifteen years old, was sent up towards the middle of August, 1868, from Warwickshire, in such a state of weakness that he could not even bear to be carried from the platform of the Great Western Railway to the Great Western Hotel without enduring much pain. He was ailing during the last eighteen months, and notwithstanding the so-called tonic treatment of several practitioners in Leamington and Cheltenham, continued to suffer and to lose power. The last six weeks before my first visit he was constantly in bed, and a short time before he took to his bed one of his legs began to drag to such an extent that he was quite unable to continue walking with his father. First I found him scarcely able to turn in bed, partly in consequence of want of power, partly from pain, of which he complained in almost every part of his body, from the head down to the hands and feet. He was very depressed; could not think well; his eyes and forehead ached; light and sound hurt him; whenever he was touched he complained of pain; he had no appetite, and took very little food; the abdomen was tender, and he suffered from constipation; passing of water caused pain, and there was soreness over the region of the bladder. Although tall for his age, he was thin; all muscles flabby; the skin, which was dry, hyperæsthesic almost all over. All these symptoms induced me to believe that onanism might be the cause of the various complaints, and of the general loss of power. My opinion was confirmed by the statement of the patient that from the age of seven, when he was first sent to a boarding school, till two months before I saw him, he had been in the habit of indulging in bad practices. He appeared not to be aware of the bad consequences, or that his complaint had some connection with those bad habits.

At my suggestion he was brought to Brighton. Vichy waters, especially les Celestines, in small quantities, French and Hungarian red wines, a mild hydropathic treatment, various manipulations on his feet and hands, milk, fruit, and vegetable diet, with substantial gravy and other nourishing soups, several hours in the open air while in a horizontal position, were the means employed during the first three months. At this time he was able to walk a short distance, and to come to my house. During the next three months he was regularly three or four times per week under the treatment by movements, had besides two or three times per week a swimming bath. Although able to walk about an hour or an hour and a half without any pain, he had still occasional attacks of depression. In February and March he was seen only eight times; and in April he had perfectly recovered the use of his limbs, and lost his previous pains. All morbid symptoms disappeared, and when his father came to see him in May, he walked two to three hours without any particular fatigue, and returned home entirely changed in appearance, in very good spirits, and able to resume his studies.

VII. *Paraplegia, with pes equinus, and one perfectly* LOOSE *hip.*

Lady —— was paralysed when five years old, after an attack of scarlatina. Since that time a club foot was developed on both sides, and tenotomy performed, and many eminent medical men having been consulted, and their treatment not proving more successful than tenotomy, she —at the age of nine—fell for two years into the hands of a female rubber in or near Liverpool, who is clever enough to persuade people that she can cure the paralysed, the deformed, and even ankylosed kyphotic curvatures. Dr. Radcliffe was consulted when she was eleven years old, and he kindly recommended my treatment. When she called the first time she was carried from one place to another, as had been done for the last six years. As a proof of what the rubber has done for her, I was shown that she could swing to and fro while supporting herself on both well-stretched arms, which were placed on two chairs. Her shoulders were much raised; the neck shortened; the spine very considerably curved—it was a double scoliosis with slight lordosis; one hip was perfectly loose—that is, all muscles round the joint paralysed, and the ligaments perfectly relaxed; both legs dangling, the one with the loose hip still more than the other; both legs, again, slightly affected with talipes equinus, notwithstanding the previous tenotomy; both legs and feet bluish red, cold, flabby, but without appearance of atrophy; the calves appeared stout, in consequence of the skin being very fat, while the muscular development was very deficient.

After a treatment continued for three years—with the exception of the summer months, according to the principles mentioned in this treatise—she was able to walk alone without any other help except the aid of one stick. A few months later an abscess in the lungs prevented the resumption of the treatment. Having happily, under the care of Sir T. Simpson and Dr. Duncan, recovered from this serious illness, there is hope that as soon as she is strong enough to continue the previous treatment, she will be able to walk even without a stick.

VIII. *Cerebral paralysis of both legs, with slight contraction of the hips (pedes equini vari), and paralysis of one arm, with contraction of the elbow.*

Master ———, four and a half years old, born in India, a very intelligent boy, squinting slightly with one eye, and affected in the manner named in the heading. Mr. Paget was consulted, and was kind enough to advise the mother to place the child under my care.

After three months, during which this boy was daily under treatment, and three times a week twice a day, which his nurse and mother assisted by doing at home, all passive movements necessary for the purpose of overcoming the rigidity of the limbs, the child had improved considerably. Six months later he was again for three months with me, when he recovered the use of his limbs to such an extent that I have not seen him since. In this case I made use of supports where the elasticity was supplied by vulcanised india rubber; but I prefer, at present, the elasticity of a steel spring, because the india rubber loses the elasticity sooner, and must be more frequently replaced.

IX. *Paraplegia, with anæsthesia of one leg, and hyperæsthesia of the other.*

Mr. ———, twenty-five years old, was attacked about two years ago gradually with paralysis of both legs, while under mercurial treatment, when he was exposed to damp; all that time he was suffering from secondary symptoms under the form of painless copper-coloured spots in various parts of the body. The right leg was the worst with regard to the power of moving, but was better with regard to sensation; the contrary state existed in the left leg. The right foot and knee were slightly contracted; the muscles of the right hip and that side of the spine more or less paralysed. Galvanism appeared to have been applied to excess. Slight exertion made him very tremulous, and when the point of the right foot involuntarily touched some object, it began to shake; and this weakness extended to the whole leg. The weakness of the spine caused a considerable stoop, flattening of the chest, and raising of the shoulder. Although aware of the desire to pass water, he was unable to retain it; constipation was a constant symptom.

At the beginning of the treatment he was unable to walk up stairs, and for the first three months was carried up. After eighty-seven séances in seven months he could, after the first three months, walk up stairs without assistance, was able to balance himself better, and could walk about easily with the aid of two sticks. At this period the treatment was interrupted, and he went to a hydropathic institution, where his mode of life was much changed, and he returned much weaker. During the two following winters he was seen twice or once a week. According to the patient's statement, Dr. Radcliffe (who had seen him before I began the treatment) expressed his surprise at the great improvement the patient had made, he being able to walk more than a mile with the aid of a stick, and to ride on horseback. The improvement continued, and now he is able to walk a short distance without the aid of any support, and without a stick.

In this case it was difficult to induce the patient to leave off his accustomed and beloved aperients, to overcome, when he was better, his fear of falling, to persuade him to walk in the open air, to swim and to ride. I am glad to say that a perfect recovery is reasonably to be expected if the patient continues his treatment.

X. *Local paralysis of one sterno-cleido-mastoideus, with slight lateral curvature.*

Miss ———, ten years old, had a slight paralytic seizure when four or five years old. Her head was inclined forward and slightly turned to the left, in consequence of the weakness of the muscles of one side of the neck. Notwithstanding tenotomy of one sterno-cleido-mastoideus performed by an orthopædic surgeon in London, and notwithstanding the treatment of a professional rubber in Liverpool, with whom she remained for a longer period, the deformity of the head continued, and a lateral curvature of the spine was developed. At Dr. Leadam's advice she was placed under my care. In the course of two years, during which the treatment was several times interrupted, she was perfectly cured. Besides the manipulations which were daily done at home, partly to diminish the action of the contracted muscles on one side, and to stimulate the nutrition and action of the weak antagonists, the treatment was principally directed at first to the straightening and strengthening of the spine, and afterwards active movements of the weak muscles of the neck, with resistance of a second person, were combined with perseverance, till the equilibrium between the muscles on both sides of the neck was entirely restored, and thus the natural position of the head obtained. Much attention was also given to her diet, which was tonic, and to her positions while at lessons. The swimming bath in tepid sea water was used twice a week.

XI. *Local paralysis, with a high degree of limping.*

Lady ——, nine years old, with the exception of chronic ozæna, a healthy, well-developed, intelligent child, has been since her infancy very lame. The cause could not be ascertained; after passing for years successively through the hands of several orthopædic and other surgeons, who did not succeed in improving her, she was entrusted, like so many others, for two years to a rubber (whose wonderful cures, performed under the auspices of a higher providence, are recorded in a book which has passed through several editions, and was favorably reviewed by English medical journals). This child limped to such an extent that the trunk and head moved at each step laterally to the left, thus forming half of an inverted pendulum movement, of which the sacro-lumbar joint was the apex. In consequence of a double lateral curvature, with slight lordosis, the length of the upper part of the body above the hip was apparently short in proportion to the length from the hip downwards. In the lying position I can scarcely observe more than half an inch difference in the length of the two legs. I am not ashamed to confess that it was only after she had been for some time under my care that I found out her incapability of raising the left leg while leaning in a standing and reclining position, and later only I observed that the power of adduction and of turning inwards the left thigh was almost entirely lost; the curvature was a secondary effect of her mode of walking. She was always projecting the right hip, and the right hip-joint formed the axis round and on which the body was moved.

The trunk was necessarily bent considerably to the left while standing on the left leg in order to enable her to move the right leg forwards. There is no doubt that infantile paralysis preceded, which localised itself in the flexors and adductors of the left thigh; the left os femoris is about three fourths of an inch shorter than the right, and this difference is made up in her sole being raised in the same proportion, in order to cause equal height of both hips. After a treatment of three years, which included bathing and swimming in tepid swimming bath and open sea, various douches, manipulations, and the movement cure, she recovered so far that she has at present a very nice figure, is able to walk an hour without fatigue, and when her attention is directed to her walk only a very slight limp is observed. As her treatment is still continued, it is probable that her best mode of walking, which at present still requires increased attention, will become habitual.

It would be tedious to give more details on the various forms of localised paralysis which cause deformities and contractions in the hip, knee, and ankle-joints, with their manifold combinations. But it may suffice to mention that the majority of these secondary effects can be prevented, even when the original paralytic affection should not be cured; the instances of paralytic complaints amongst the young are very rare, in which a suitable treatment, even after the invasion of the disease, continued with perseverance, does not cause a more or less considerable amount of improvement.

www.ingramcontent.com/pod-product-compliance
Lightning Source LLC
Chambersburg PA
CBHW020112170426
43199CB00009B/504